CHARLOTTE KLONK (ED.)
NEW LABORATORIES

NEW LABORA-TORIES

HISTORICAL AND CRITICAL PERSPECTIVES ON CONTEMPORARY DEVELOPMENTS

EDITED BY CHARLOTTE KLONK

This publication was made possible by *Image Knowledge Gestaltung. An Interdisciplinary Laboratory Cluster of Excellence at the Humboldt-Universität zu Berlin* with financial support from the German Research Foundation as part of the Excellence Initiative.

Bild
Wissen
Gestaltung
Ein Interdisziplinäres Labor

Gefördert durch die
DFG

ISBN 978-3-11-037261-8
e-ISBN (PDF) 978-3-11-036693-8
e-ISBN (EPUB) 978-3-11-039248-7

Library of Congress Cataloging-in-Publication Data
A CIP catalog record for this book has been applied for at the Library of Congress.

Bibliographic information published by the Deutsche Nationalbibliothek
The Deutsche Nationalbibliothek lists this publication in the Deutsche Nationalbibliografie;
detailed bibliographic data are available on the Internet at http://dnb.dnb.de.

Cover illustration: Menno Aden
Design: Katja Gretzinger
Typesetting: Rüdiger Kern (Berlin)
Translation: J. M. Denton (Berlin)
Printing and binding: Beltz Bad Langensalza GmbH, Bad Langensalza
Printed on acid-free paper
Printed in Germany

www.degruyter.com

"In my view, whether or not a building is aesthetically appropriate is of no great significance for the rate of productivity in scientific work. I would even say that I personally had my most important work phases in laboratories that were already rather old."

Norbert Koch, Department of Physics,
Humboldt University Berlin

1

INTRODUCTION

Charlotte Klonk

For well over a century now, the laboratory has been the epitome of creative experimentation, demonised and glorified as such in equal measure. Cinema since the silent film era has been rich in images of the genie run wild, pursuing his creative urges in solitary places, at the far side of good and evil—Rotwang in Fritz Lang's *Metropolis* (1927) is perhaps the most famous example. The realms of architecture, art and museology, by contrast, were quick to recognise and lay claim to the more positive connotations. Two years before *Metropolis*, the architect Walter Gropius spoke of the workshops at the newly founded Bauhaus as laboratories (Gropius and Moholy-Nagy 1925), and in 1939 Alfred H. Barr, the Founding Director of the Museum of Modern Art in New York, declared it a laboratory for the arts (Barr 1939). But what image of the laboratory do we have in mind when we stylise it as the *locus classicus* of the inquiring and inventive intellect?

To fathom this question, the authors of the present volume have journeyed into the past and present of laboratory architecture. Firstly, they asked which images of the research scientist and the laboratory have been purveyed since the natural sciences were institutionalised in the nineteenth century. Secondly, they examined contemporary architectural solutions in the light of influential laboratory architectures of the latter half of the twentieth century, and thereby discovered a great variety of historical and contemporary approaches—for both the functional interrelation of spaces and the tension between symbolic façades and internal structures can take very different forms.

So, is there any such thing as a typical twenty-first-century laboratory given these variables? Put the question to natural scientists and one usually receives a straightforward answer: architecture is a secondary issue and hence accorded little importance in the research context. The quotations that precede each chapter overwhelmingly substantiate this view. They are taken from interviews that the authors conducted with natural scientists in Europe, the United States and Asia. Although the interviews cannot claim to be representative in the empirical sense, they confirm that research scientists' sense of belonging to international networks of experts generally carries much more weight than any one physical workplace. The interviewees were almost unanimous on one point, however—namely that changes in laboratory architecture reflect the changing face of scientific research.

According to the historian of science Peter Galison, three distinct phases can be discerned in the twentieth century:

> In many ways prewar scientists [...] saw themselves as essentially solitary in their confrontation with nature. By the mid-1940s, that isolated self-understanding, previously reinforced by small-scale labs and studios, began to break down as [...] scientists [...] encountered patterns of work and places of production altered by the vast factory production quotas of firms like Albert Kahn Associates and the Austin Company. Physical plants, social ordering, and a new subject position entered together in the government-sponsored factory-laboratory exemplified by Oak Ridge, Hanford, and Los Alamos.
> (Galison 1999, 18)

According to Galison and his co-author Caroline Jones, these industrial-style, hierarchically organised laboratories have gradually ceded place to a new model of research that emerged in the 1970s. Instead of standing at machines, scientists now spend more time at their computers; instead of working exclusively in centralised institutions, they increasingly pursue decentralised research projects where "experiments are dispersed social-technical-spatial entities in which meaning is constructed at several peripheries" (Galison and Jones 1999, 527). Architecturally, this is evinced, for example, by the structure of the European Centre for Nuclear Research (CERN) in Switzerland, which has a gigantic underground central experimentation tract yet transmits its data to hundreds of digitally networked laboratories all over the globe. This obviates the need for a particularly representative architecture—and the complex does indeed comprise merely a straggle of buildings erected ad hoc on the site over several decades (Galison 1999, 18–19).

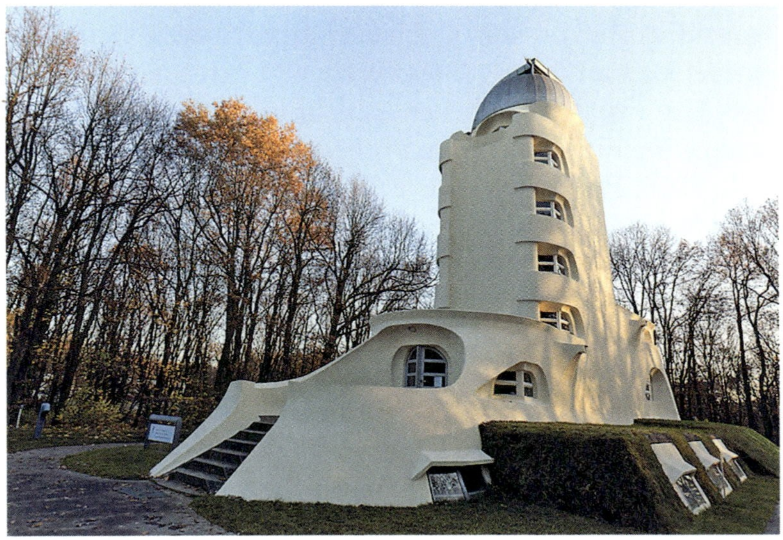

Fig. 1.1:
Erich Mendelsohn, *Einstein Tower*, Potsdam, 1919-22. Photo © Dario J. Lagana.

When Galison wrote this in 1999 it was impossible to foresee that a boom in laboratory architecture would set in at the turn of the millennium. Since then, architects hitherto renowned for museum projects—Rafael Moneo, Yoshio Taniguchi, and David Chipperfield, to name but a few—have built iconic research centres of international repute. Public appreciation has grown apace. In 2013 alone, the arts section of the German daily *Frankfurter Allgemeine Zeitung* twice featured reviews of new laboratory buildings (Grünzig 2013, 38; Hamm 2013, 29) while the *Weekend Magazine* of the UK daily *Financial Times* published an extensive article on the subject by the architecture critic Edwin Heathcote under the title "Cathedrals of the Mind" (Heathcote 2013, 38–43).

Yet "signature buildings", so called because their form bears the unmistakeable stamp of an architect's style, are nothing new in the field of laboratory architecture. As early as 1919–22, the architect Erich Mendelsohn gave expression to Albert Einstein's theory of relativity—his new notion of a time-space curve—in his design of a modern laboratory and observatory building in Potsdam, Germany. The result is unique yet nonetheless typical of Mendelsohn's architecture (**Fig. 1.1**): with no right angle in sight, the impression is of space distorted by heavy objects in the Tower's vicinity or of a *blob* extracted from the earth and vertically extended. Contemporary architects likewise sometimes endeavour to express in the façade of a laboratory building the nature of the scientific research pursued within. In 2010, Professors Andre Geim and Konstantin Novoselev of the University of Manchester in the United Kingdom were jointly awarded the Nobel Prize for their discovery of the material graphene. In March 2015, the University opened the new National Graphene Institute, thereby literally building on its research scientists' resounding success. Jestico + Whiles, the architects of the project, clad the entire façade in translucent dark stainless steel, in reference to the qualities of the material discovered by Geim and Novoselev (**Fig. 1.2**).

However, buildings do not generally become world famous because architects identify a specific form apt to express the research subjects pursued within them, but rather because architects develop their own spatial and formal idioms. One early example of such iconic laboratory architecture is the Salk Institute for Biological Studies in La Jolla, California, designed in the 1960s by Louis I. Kahn (**Fig. 2.1** [See Chapter 2]); another, the Lewis Thomas Laboratory for Molecular Biology, opened on the Princeton University campus in 1986 (See Chapters 4 and 5). Robert Venturi and Denise Scott Brown designed the latter's exterior; James Collins, Jr. of Payette Associates its interior (**Fig. 5.11** [See Chapter 5]).

The Lewis Thomas Laboratory served Peter Galison and Emily Thompson as a case study in their co-edited book, *The Architecture of Science*, even though it does not fit their description of the development of laboratory architecture in the twentieth century. Venturi consciously avoided in his façade design any sign that might betray the building's purpose. Instead he chose, as is typical for him, to cite vernacular formal elements, which in this case evoke among other things a traditional New England country house. The interior of the building, by contrast, suggests a solidly built school. It has four separate research units linked by restricted access staircases and discreet

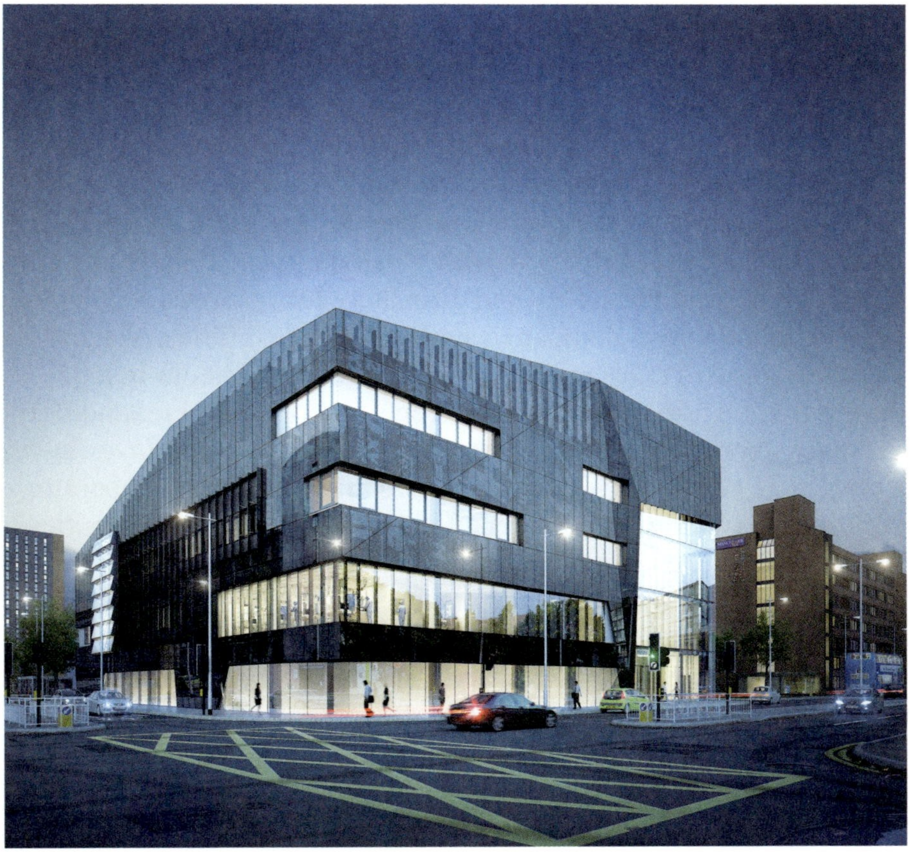

Fig. 1.2:
Jestico + Whiles, *National Graphene Institute*, Manchester, 2015.
Design © Jestico + Whiles.

conference rooms. Given its singular style, the laboratory is neither a factory-like structure nor an architecturally nondescript complex of buildings and experimentation tracts erected in the service of a globally networked yet no longer locally present research community. Instead it constitutes an idiosyncratic, custom-made solution to the architectural task at hand.

Galison does, however, mention a substantial shift that occurred in the late twentieth century, first in the United States and then in Europe, one that now retrospectively explains the genesis of the Lewis Thomas Laboratory. Its distinctive architecture heralded a new trend, the repercussions of which only really made themselves felt by the turn of the millennium: "the new laboratory design also reflected the changed realities of funding. [...] the shift from foundational to government funding entailed the need to [architecturally] support spaces that would allow the kind of interdisciplinary collaboration that the government wanted to support" (Galison 1999, 15). Moreover, the new entrepreneurial ethos that came into play in the 1990s increasingly compelled academics to orient their research to concrete applications and technological feasibility and ultimately led to closer collaboration between commercial companies and research institutions (idem). This trend contributed substantially to the international boom in laboratory architecture over the last fifteen years or so. In England, for instance, approval of the state-funded *Science Research Investment Fund* (SRIF) in 2002, which earmarked millions for new laboratory buildings, temporarily put an end to a period in which universities had struggled to make ends meet; and in Germany an *Excellence Initiative for Cutting-Edge Research* was launched in 2006, and the Federal and State Governments reached an administrative agreement the following year on the necessity of organising a Joint Science Conference. One of the case studies discussed in the present volume—the new research building foreseen for the project "Hybrid Systems for Electronics, Optoelectronics and Photonics" at the Humboldt University Berlin—was approved at this conference in 2013 and facilitated by the new funding structure (See Chapter 8). As is the case with government-funded architectural projects, forging links between interdisciplinary foundational research and application-oriented knowledge is of crucial importance. Accordingly, the idea is to bring together chemists, physicists, theoreticians and experimenters, also from the private sector, to collaborate on new hybrid materials.

Interdisciplinary tie-in research of this kind is generally a reason and a stimulus for the construction of new laboratories. It also demands a new laboratory of the sort that Oliver Milton of Hawkins/Brown, the architects of the Biochemistry Laboratory and the new physics building at Oxford University, says is primarily defined by the need for spaces of communication: "The key driver behind these buildings is collaboration. [...] So you need to create an environment which allows [scientists] to do their own stuff but then also come together. So, while you make the offices and labs as good as possible, the building is really all about the spaces in-between" (quoted in Heathcote 2013, 42). After the Second World War, architecture was required to accommodate complex technologies and large machines, but, by the turn of the millennium, attention shifted to the design of connective spaces and points of encounter located in-between the offices and laboratories of research staff (See Chapter 6). The emphasis is now on informal contact that will give rise to interdisciplinary exchange hitherto thwarted by a mono-disciplinary focus and the structural separation of different academic departments. Thus, in contrast to Peter Galison and Caroline Jones's assumption that the trend in the natural sciences was to decentralised virtual networks (Galison and Jones 1999, 524), the last ten years at the least have witnessed a marked resurgence of the research laboratory as a physical place.

To foster interdisciplinary and possibly also spontaneous communication is now widely acknowledged to be the goal of any new laboratory building. Yet the ability to attain that goal varies enormously, depending on budget capacity. A private company, the Novartis AG, commissioned an entire "laboratory city" in the St. Johann district of Basel. Each building was designed by a different renowned architect and hence carried his or her specific signature. Moreover, all the architects were asked to pay particular attention to spaces of connectivity and communication (See Chapters 4 and 5). Private universities in the United States too—MIT, Princeton, and Harvard for example—have likewise been commissioning famous architects for some time now, in order to create distinctive campuses and gain an edge in the global competition for leading scientists. However, exclusively state-funded construction projects generally cannot afford iconic statements of this kind. Here, the commission often goes to architect offices specialised in laboratory architecture or, alternatively, to relatively unknown firms with whom the client may or may not strike lucky.

The University of Oxford, for example, in a closed architectural competition for its new Biochemistry Laboratory completed in

2009, commissioned the team of Hawkins/Brown at a time when the firm was not yet well known for its laboratory architecture. The laboratories and offices are clustered around a distinctively designed atrium that provides users with both communal spaces and quiet retreats (**Fig. 1.3**). Although the building has won many awards since its completion and much praise from its users, it is not free of the

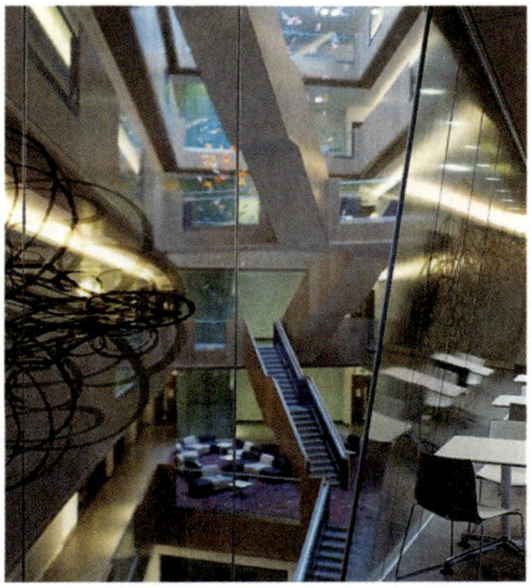

Fig. 1.3:
Hawkins/Brown, *Biochemistry Laboratory*, Oxford, 2009. Photo © Hawkins/Brown.

compromises so typical of state-funded projects. As the funding was specifically earmarked for a research building, it could not be used to build lecture halls and seminar rooms. Such terms and conditions ultimately make it impossible to uphold the union of research and teaching at universities that once propelled the proliferation of laboratory buildings in the nineteenth and early twentieth centuries (See Chapter 7).

The new research building IRIS (Integrative Research Institute for the Sciences) at the Humboldt University Berlin has also been plagued since the planning stage by unavoidable constraints. Although the prospective users in collaboration with the authors of the present volume had developed a very clear idea of the significance and necessary sequence of informal spaces in the building, their proposals were integrated in the plans only after the Berlin Senate for Urban Development and the Environment (the official client) had held an architectural competition and commissioned a firm to realise the project. Moreover, each and every one of the proposals ultimately fell victim to funding cuts (See Chapter 8). However ingenious planners and designers may be in such situations, it is clear at the end of the day that the various terms and conditions in place, be they financial, technical or due to urban planning regulations, basically foster one thing only: the generic laboratory. As a rule, nowadays, this means a generous atrium (often with a cafeteria), flanked on one side by laboratories, on the other by offices and meeting rooms. Glazing too is now more or less standard, so as to ensure that all spaces are visually connected, transparent and open, even though this may at times be inappropriate (See Chapter 4). To name just one instance, a glazed lecture hall can never be made dark enough to show fluorescent images as successfully as it can less light-sensitive images. Payette Associates, an American company specialised in laboratory construction, developed the prototype for this laboratory in the 1990s. The company has since won many awards, most recently for the Frick Chemistry Laboratory on the Princeton campus, which it co-designed with Hopkins Architects and completed in 2011 (**Fig. 5.13** [See Chapter 5]). Eventual modifications notwithstanding, this generic type has now set the benchmark for laboratory architecture in the twenty-first century.

What does this imply for the public image of both the laboratory and the research scientists who work there? Primarily this: the role model today is no longer the secluded space or the factory but rather the totally upbeat, state-of-the-art media and communications company. As Oliver Milton of Hawkins/Brown explained, "when we are looking for references, it's often more useful to look at ad agencies or tech companies like Google or Apple" (quoted in Heathcote 2013, 32). In the twenty-first century, laboratories are centres of communication and hence those conducting research within them are now cast as communication experts. This, according to the authors of this book, is the new image of the natural scientist as reflected in architecture since the turn of the millennium. So far this development has not been studied in depth. Since the new discipline of laboratory studies was established in the course of Bruno Latour's and Karin Knorr Cetina's field investigations in the 1970s, the main issue has been to examine the laboratory as an epistemological complex and as a social system with its own cultural codes. In this volume, **Christina Landbrecht** and **Verena Straub** give an account of the development of laboratory studies, juxtaposing sociological and historical approaches. The laboratory, they argue, is a space that has always accommodated a variety of practices and as such has no fixed connotation. Its meaning changes, they conclude, as research in and about it changes.

A study of the iconography of the laboratory has been a desideratum for some time. In *The Image of the Laboratory*, **Verena Straub** first examines photographs of the laboratory from around 1900. She finds not only a modern interpretation of the solitary research genius, such as Fritz Lang took up in *Metropolis*, but also conventionalised visual formulae, for example the scientist dressed in a white coat and surrounded, significantly, by laboratory equipment, even when none of it in any way relates to his (or, more rarely, her) work. Aesthetic, stylised images of this kind still resonate in Hollywood films and shape contemporary notions of the laboratory, Straub concludes. **Christina Landbrecht** turns in the following essay to the images that architects and those who commission laboratories have been circulating since the 1960s. Transparency, as she demonstrates, has become a major objective. Like public accessibility, it was initially a criterion only for a few individual clients, owing to their humanist or political convictions. Today, however, the installation of large amounts of glazing has become standard. Yet the demand for transparency propelling this trend, Landbrecht argues, is mostly nothing more than a PR ploy; only very rarely does it signify actual public access or the right to monitor work carried out in the laboratory.

In *Function and Design*, the focus is not on the image but on the architecture of the laboratory. **Sabine Hansmann** draws on pertinent examples—Louis I. Kahn's Medical Research Laboratory in Pennsylvania; Venturi, Rauch and Scott Brown's Lewis Thomas Laboratory in Princeton; and Hopkins Architects and Payette Associates' Frick Chemistry Laboratory, likewise in Princeton—to demonstrate the virtually paradoxical requirements that architects face. They are briefed, on the one hand, to create highly flexible, modular spaces that can be easily adapted to the needs of various new research projects and, on the other, to design buildings that have something special or even unique to offer and are hence likely to forge a sense of community among research scientists and to lend the institution an iconic profile. Analysis of laboratory buildings in terms of their function, such as **Henrike Rabe** provides in the following essay, shows that the distribution of informal spaces varies widely from place to place, but that their floor area has increased significantly, overall, in recent years. Given the current focus on communication it is wholly conceivable that architects in the future may find some not yet exploited opportunities here.

In the final section, *Berlin Case Studies*, two natural scientists describe the evolution of Berlin's chemistry and physics laboratories from the late nineteenth century to the present day. **Ronald Göbel** explores the history of the Institute of Chemistry and thereby demonstrates that the building's structure and usage as well as the academic themes successively pursued within it have continually shaped one another. His correlation of ground plans and teaching curricula reveals that teaching and research were accorded more or less equal importance in the institution's early years, although teaching had a slight lead in terms of the space allotted to it. Its importance increasingly diminished in subsequent decades, however, owing to the upheaval of two world wars and the division then reunification of Germany. Today, far more space is devoted to research than to teaching. As Göbel shows, advances in technology as well as the changing demands made on tertiary chemistry education determined both the architecture of the Institute and its curricula. Finally, **Jürgen P. Rabe** reports on the planning of the new research building IRIS foreseen for the project "Hybrid Systems for Electronics, Optoelectronics and Photonics" at the Humboldt University Berlin. He draws on his personal involvement as Founding Chair and prospective user of the facility and argues that, while in some respects the new laboratory perpetuates the traditions of its late-nineteenth-century predecessors, it also marks a departure by actively fostering interdisciplinary tie-in research.

"Buildings serve both as active agents in the transformation of scientific identity and as evidence [of] these changes," stated Peter Galison back in 1999 (Galison 1999, 3). In the meantime there are hopes that architecture may also contribute substantially to generating as yet unforeseeable innovations. Whether such optimism is warranted is an open question. The pertinent research has yet to be done (Bonetta 2003).[1] This book, with its reflection on images and its analyses of current conditions, can aspire to do no more than provide helpful orientation, not only for cultural and social historians but also for architects and expert planners, and for those who commission laboratories as well as those who use them. The challenges are still considerable. A gigantic laboratory building—the Francis Crick Institute—is currently under construction in the booming London district of King's Cross and will eventually house six hitherto independent leading British cancer research institutions under one roof. Here too, the goal is to create a place of interdisciplinary encounter and exchange for the more than 1,500 people who will work there. How this goal might effectively be attained for such a large number remains to be seen, but for the moment at least, the focus on communication is common currency.

REFERENCES

Allan, Thomas J. 1977. *Managing the Flow of Technology*. Cambridge, MA: MIT Press.

Barr, Alfred H. 1939. "Art in Our Time: The Plan of the Exhibition." In *Art in Our Time: An Exhibition to Celebrate the Tenth Anniversary of the Museum of Modern Art and the Opening of Its New Building*, 13-15. New York: Museum of Modern Art.

Bonetta, Laura. 2003. "Do you want to work here?" *Nature* 424: 718-20.

Galison, Peter. 1999. "Buildings and the Subject of Science." In *The Architecture of Science*, edited by Peter Galison and Emily Thompson, 1-25. Cambridge, MA: MIT Press.

Galison, Peter and Caroline A. Jones. 1999. "Factory, Laboratory, Studio: Dispersing Sites of Production." In *The Architecture of Science*, edited by Peter Galison and Emily Thompson, 497-540. Cambridge, MA: MIT Press.

Gropius, Walter and László Moholy-Nagy. 1925. *Bauhausbücher 7: Neue Arbeiten der Bauhauswerkstätten*. Munich: Albert Langen Verlag.

Grünzig, Matthias. 2013. "Das Glas der Weisen." *Frankfurter Allgemeine Zeitung*, July 13.

Hamm, Oliver G. 2013. "Das Pantheon für Röntgenlaser." *Frankfurter Allgemeine Zeitung*, August 13.

Heathcote, Edwin. 2013. "Cathedrals of the Mind." *Financial Times Weekend Magazine*, October 19/20.

Hillier, Bill and Alan Penn. 1991. "Visible Colleges: Structure and Randomness in the Place of Discovery." *Science in Context* 4 (1): 23-49.

1 The first studies of the interrelation of spatial distribution in laboratories and their users' communication patterns were published in the 1970s (Allen 1977). Ten years later, Bill Hillier and Alan Penn drew on these in their analysis of the spatial conditions in two laboratories whose respective teams clearly differed in terms of their communication patterns, research activity and output (Hillier and Penn 1991). However, Hillier and Penn could not ascertain the degree of innovation and "research productivity" attained in the labs they investigated (idem, 45).

"Research labs have been globalised or 'uniformed' in the same way that shopping centres look exactly alike, no matter where you go."

Antonio Castro Neto, Graphene Research Centre, National University of Singapore

2

THE LABORATORY AS A SUBJECT OF RESEARCH

Christina Landbrecht and Verena Straub

The laboratory is the linchpin of scientific experimental practice. One major German reference work describes it as a workplace for natural scientists that is equipped with the requisite appliances "for scientific and technical experiment and measurement, and tasks of evaluation [and] monitoring" (Brockhaus 2006). The *Oxford English Dictionary* goes a step further, defining the laboratory as a building type that may serve not only for research but also for production or instruction: "[A laboratory is a] room or building equipped for scientific experiments, research, or teaching, or the manufacture of drugs or chemicals" (OED 2014). Yet a quick glance at the history of research into the laboratory since the 1970s suffices to show that the question "What is a laboratory?" can be answered in a number of ways. In this chapter, we first discuss the development of sociological research into the laboratory and then turn to the contribution made in recent years by historians of science who have begun to investigate the concrete spatial and architectonic settings of scientific research.

Unlike encyclopaedias, the sociology of science of the late 1970s did not define the lab purely in terms of its purpose and spatiality but turned the spotlight instead on its epistemological function. Until then, sociologists had barely addressed everyday working procedures in the laboratory. Rather, their focus had been to analyse the macrostructures of natural scientific research. This involved investigation of the dissemination of findings (Frame et al. 1977), of funding structures (Korach 1964) and of the interplay of science and politics, including the mechanisms by which research projects were approved (Gilpin and Wright 1964). There were exceptions, however. In their study *Laboratory Life,*

Bruno Latour and Steve Woolgar cite the sociologist of science David O. Edge, who, in a lecture at the University of Berkeley in 1976, had described his early research interest in (everyday) conversations among scientists as a means to examine what he called "the soft underbelly of science" (quoted in Latour and Woolgar 1986, 27).

Soon afterwards this "soft underbelly of science" came to constitute a whole new branch within sociology, namely the sociology of science, prompting sociologists and anthropologists to begin analysing working procedures in the laboratory. Such early field research-based ethnographies of the laboratory—which were propelled from the mid- to late 1970s by the author duo Latour and Woolgar, the sociologists Karin Knorr Cetina (Knorr Cetina 1981) and Michael Lynch (Lynch 1985), and the anthropologist Sharon Traweek (Traweek 1988)—are hailed today as milestones in the history of laboratory studies and also contributed substantially to establishing the latter as a sub-discipline of Science and Technology Studies (STS). The primary focus of laboratory studies was the analysis of working practices in the laboratory and the attendant issue of how scientific findings were constructed and disseminated. The methodology was based on participant observation—the typical means of data collection in ethnology—and complemented by elements of discursive analysis. Thus it was possible to examine the production of scientific facts as a constructive (as opposed to descriptive) process.

This perspective marked an entirely new departure for sociology, since the assumption that facts were incontrovertible entities had persisted unchallenged until well into the 1970s. For Knorr Cetina, however:

> Constructionism is one of the major, perhaps the major, outcome of laboratory studies [...]; the origin of its emphatic use in STS lies in the attempt of students of laboratories to come to grips with their observational record of the 'made' and accomplished character of technical effects.
> (Knorr Cetina 1995, 141)

The findings of early laboratory studies were based on data that sociologists had compiled during months of field research in various laboratories. Latour, for example, held a part-time job from October 1975 to August 1977 at the Salk Institute in California. He was employed as a technical assistant in the laboratory of medical research scientist and later Nobel Laureate Roger Guillemin, who was a major figure in the field of neuroendocrinology. In contrast to Latour's focus on a specific laboratory and its working practices, Knorr Cetina's concurrent research was broader in scope. Her analysis at the University of Berkeley, conducted from October 1976 to October 1977, was based on in-depth studies of an entire research centre comprised of seventeen different research units in which both foundational and applied research were pursued:

> Two of these units were devoted to chemistry while others dealt in plant biochemistry, plant photochemistry, toxicology, microbiology, chemical analysis, instrumental analysis, fibre science and food technology. Two units worked in the field of food engineering, with the other six being orientated more toward general problems than specific disciplines.
> (Knorr Cetina 1981, 24–25)

Knorr Cetina's approach differed from Latour's in that she portrayed the laboratory not as a concrete space but as an operational unit. While Latour in *Laboratory Life* described Guillemin's laboratory in terms of its spatial set-up and also included in his book a sketch of the ground plan, Knorr Cetina in *The Manufacture of Knowledge* painted an abstract image of the laboratory. She sought to present a theoretical conception that could be understood as an epistemological construct, emphasising that the laboratory itself limits the scope and the form of scientists' activity yet is nonetheless also significant as a physical place. Within this framework she was able not only to observe and analyse the process of scientific knowledge production but also to describe scientists' activity as a culture apt to be researched using the ethnologist's methods and means. Thus, for Knorr Cetina, the laboratory was also always more than *the locus* of scientific activity. To her, it represented the elementary range of tools by which knowledge of the natural sciences might be attained. It is, she stated, "to use a term of Bourdieu, […] the locus of mechanisms and processes that help explain and guarantee the success of the modern natural sciences" (Knorr Cetina 2002a, XIV).

The laboratory came to signify something more than the concrete site of actual research. In Knorr Cetina's analyses, the topos "laboratory" retained its relationship with the world, since scientists worked there on subjects of research that were derived from nature. Yet it simultaneously constituted an "improvement" on the natural environment:

> A laboratory science [must] not inquire into objects of knowledge there, *where they happen to be*, i.e. wherever they are anchored in their natural environment; […] nor must it take these objects *as they are,* since it can substitute them with all manner of reduced and partial objects. Finally, a laboratory science must not inquire into things *when they happen*; it can disengage them for example from natural cycles, such as periodic recurrence and growth.
> (Knorr Cetina 2002a, XIV)

According to Knorr Cetina, dealing with nature in the laboratory context amounts to a system of "reconfiguration" (Knorr Cetina 1995, 145). This idea owed much to the French phenomenologist Merleau-Ponty, who wrote "that classical science is a form of perception which loses sight of its origins and believes itself complete" (Merleau-Ponty 1962, 57). Merleau-Ponty accordingly posited that "our task will be […] to rediscover phenomena, the layer of living experience through which other people and things are first given to us, 'Self-others-things' as it comes into being" (idem). Accordingly Knorr Cetina claimed that not natural phenomena but only "images of such phenomena" were processed in the laboratory (Knorr Cetina 2002a, XV). The laboratory is hence, on the one hand, a mechanism that modifies and manipulates natural objects, and, on the other, itself always evolving as a "laboratope" (Knorr Cetina 2002b, 58), an artificial ecosystem that both influences the natural order of things and reconfigures the social order—for the agents of laboratory research are as subject to transformative processes as the objects of their research are:

> If we see laboratory processes as processes that 'align' the natural order with the social order by creating reconfigured, 'workable' objects in relation to agents of a given time and place, we also have to see how laboratories install 'reconfigured' scientists who become 'workable' [feasible] in relation to these objects.
> (Knorr Cetina 1995, 146)

This notion of the lab as a distinctive place enormously influenced the mindset of early laboratory ethnologists, as is evident from Latour's comments on the Salk Institute.[1] From 1959 to 1965, the architect Louis I. Kahn designed the Salk Institute in close consultation with its Founding Director, the virologist Jonas Salk. The result is an outstanding, solitary structure, perched on a clifftop above the Pacific Ocean, on the outskirts of Los Angeles (**Fig. 2.1**). Despite initial funding problems, its renown as a research centre for molecular biology and genetics spread internationally within only a few years and brought together leading scientists, among them the Nobel Laureates Francis Crick, Roger Guillemin and Renato Dulbecco (Leslie 2005, 167–68). In his notes taken during field research at the Salk Institute—which were published only in 1988, in the French translation of *Laboratory Life*—Latour describes the Institute as "a mix of Greek temple and mausoleum" (Latour 1988, 11), i.e. as a place remote in terms both of its location and its architecture and thus consonant with the popular image of the laboratory.

Sharon Traweek, in her analysis of the SLAC research institute (Stanford Linear Accelerator Center) in California and the KEK (Ko-Enerugie butsurigaku Kenkyusho) in Japan, likewise notes that the laboratory is marked as a distinct place. She demonstrates this fact by reference not only to its actual structure—which is

1 It is surprising to note that Latour made no mention of the Salk Institute's renown in the English edition of *Laboratory Life*, all the more so given the two reasons which, as his field notes reveal, decisively shaped his choice of this research location: firstly, because Guillemin—who was not only a fellow Frenchman but also hailed like Latour from Burgundy—approved his research proposal; and, secondly, because the Institute had a certain prestige: "This wealthy and famous laboratory was quite a welcome contrast to the Ivorian employees I had just finished studying" (Latour 1988, 11)

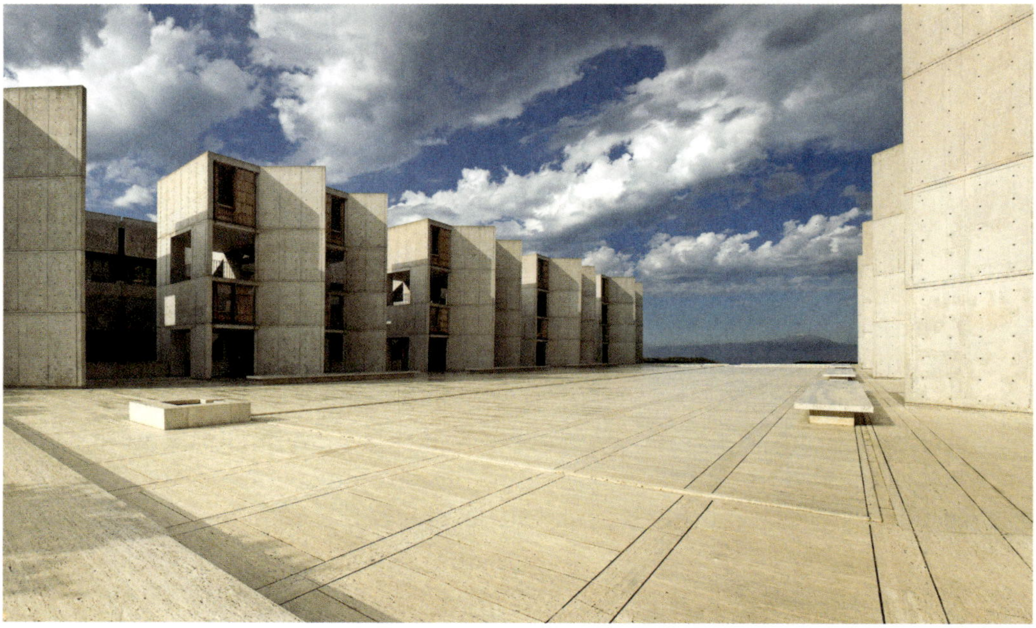

Fig. 2.1:
Louis I. Kahn, *View of the Courtyard*, The Salk Institute, La Jolla, California, 1959-65.
Photo © Joe Belcovson.

often dictated by security concerns—but equally to perceptions firmly anchored in the minds of the general public.

> The SLAC radiation fence is a tangible border [...]. It is a symbol of the lab's restraint [and] it is also a symbol of the lab's very great and dangerous power. [...] There are no fences around the major high-energy physics laboratory in Japan [...]. It is [however] part of a new 'science city,' Tsukuba, which includes many scientific research labs, a university, and housing for the staff of all these institutions and their families. Many Japanese people say that Tsukuba Science City is 'not really Japan'.
> (Traweek 1988, 40)

Analyses such as these substantiate Knorr Cetina's description of the laboratory as a place with its own social and cultural codes, positioned in relation to its environment yet at the same time hermetically sealed. This reading of the laboratory subsequently became the unchallenged frame of reference for laboratory studies.

In order to fully comprehend the laboratory's significance as the frame of reference within which scientific facts are generated, laboratory studies devoted itself, inter alia, to the analysis of agricultural studies or astronomy, branches of science that had evolved from purely field research-based disciplines into laboratory sciences. Such analysis demonstrated that modification of the research context has consequences for the objects of research (Busch et al. 1991; Smith and Tatarewicz 1985). A further focus here—on the premise that the laboratory constitutes an "other" environment—was the specificity of processes within the laboratory: scientists' specific remarks and rhetoric were analysed (Knorr Cetina 1995, 155) along with the techniques of representation they applied in visual design and image processing (Hendersen 1999).

Portrayals of the laboratory as a place apart remain widespread to this day in the work of historians and sociologists of science. For example, the American historian of science Robert E. Kohler describes the laboratory as a mono-functional space and a neutral stage. In his historical study of the biological field research methods in use from 1880 to 1950, he claims that the lab is as much moulded by social homogeneity as the actions unfolding within it are subject to a universally valid grammar. He describes the laboratory space as a heterotopia:[2] "Labs are separate, a world apart from the world; [...] It is precisely the stripped-down simplicity and invariability of labs—their placelessness—that gives them their credibility" (Kohler 2002, 7). Labs, according to Kohler, are generic spaces that nourish the illusion that any facts produced within them are universally valid. The American sociologist Thomas F. Gieryn corroborated this thesis with his case study of the James H. Clark Center for Biomedical Engineering

2 Although the term "heterotopia" coined in 1967 by Michel Foucault has never played a role in theoretical laboratory discourse, it has a certain relevance to the portrayal of the lab as a place apart. The historian of science Robert E. Kohler used the laboratory in this sense, as a counterpoint to the field researcher's working environment. He cast the laboratory as a socially established yet privileged place based on a system of opening and closing—and hence consonant with the fifth principle of heterotypology, as defined by Foucault in "Of Other Spaces, Heterotopias" in 1967 (Foucault 1992, 44).

and Sciences at Stanford University (**Fig. 2.2**). In his view, it represents a similarly "placeless place"—just as the infrastructure of airports, corporate parks and shopping malls in the age of global capitalism does. Sociologists such as Anthony Giddens and David Harvey have been developing a theory of placelessness since the 1990s (Gieryn 2008). Gieryn pursues a similar theoretical

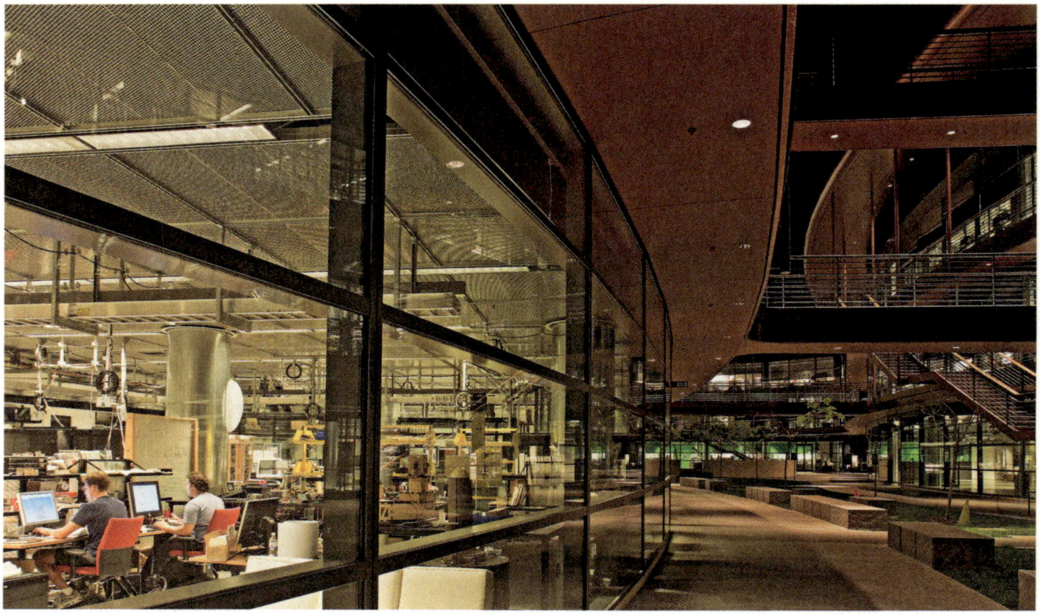

Fig. 2.2:
Foster & Partners in collaboration with Perkins & Will (formerly MBT Architecture), *View into the laboratories*, The Clark Center, Stanford, California, 1999-2003. Photo © QT Luong.

line, sketching the Clark Center—built from 2001 to 2003—as a harbinger of the twenty-first-century laboratory, neither the exterior nor interior design of which is permitted to betray even a hint of local colour. Unlike the Lewis Thomas Laboratory, for which the American architects Robert Venturi and Denise Scott Brown designed a façade that cites its neighbours on the Princeton University campus (See Chapter 5), the Clark Center has not only a façade that in no way references the surrounding university buildings but also an interior design and fittings—the latter, including laboratory workbenches, all on wheels—which perpetuate the myth of the laboratory as a generic space in which universally valid facts are brought forth. Gieryn saw this gesture as decisive for the contemporary aesthetics of laboratory buildings. He cites Kohler, who, in recognition of the fact that scientific credibility is closely tied up with the laboratory as a specific spatial type, had reached the following conclusion:

> Generic places sustain the illusion that their inhabitants' beliefs and practices are everyone's beliefs and practices. We credit knowledge and practices that are universal and mistrust what is merely local and particular, and laboratories are meant to be universal, the same everywhere.
> (Kohler 2002, 8)

In fact, however, empirical architecture plays a pivotal role not least in shaping an understanding of changes in the world of science. Knorr Cetina herself noted in view of the narrow reading of the laboratory within laboratory studies that laboratories must be understood "as places for the production of knowledge [in the light of] historical and contemporary structural variations" (Knorr Cetina 2002a, XVIII). The focus of early laboratory studies, she continued, had been limited to the traditional workbench laboratory and resulted in a far too narrow concept of the lab. Knorr Cetina therefore advocated that the scope of laboratory research be broadened to encompass historical analyses of the first domestic labs in seventeenth-century England as well as the latest developments, such as globally networked "collaboratories", for the latter type of network, now increasingly common in the fields of theoretical physics, astronomy and astrophysics, is as valid a subject of research as the traditional tabletop lab (idem).

But despite her insistence, studies devoted to the historical locus of the laboratory and its functions are still comparatively rare. "It is interesting," the historian of science Robert Anderson writes, "that in recent times laboratories seem to have come to be of greater interest to sociologists than to historians" (Anderson 2013, 671). Yet the question of *what* the laboratory is necessarily leads to the question of *where* the laboratory is located. If, however, one looks for research on the specificity and development of laboratory buildings from a historical perspective, one finds only occasional case studies, still too few in number to amount to a comparative history of the laboratory.[3] Yet as much as these historical soundings may differ, they all aspire to challenge the definition of the laboratory as "an autonomous entity", such as sociological laboratory studies proffer.

Nonetheless, the 1989 edition of the *Oxford English Dictionary* defined the laboratory as a "building set apart".[4] The historian of science Graeme Gooday challenged this definition in an article published in 2008, "Placing or Replacing the Laboratory in the History of Science?" Here, Gooday resolutely called for a history of the laboratory that encompasses the domestic or layperson's laboratory just as it does the iconic laboratories of grand institutions (Gooday 2008). This kind of inclusive history of the laboratory would have to take into account the fact that experiments were carried out not only in "buildings set apart" for that specific purpose but also in private residences, kitchens, museums, workshops, hospitals and pubs:[5]

> Rather than taking conveniently for granted the autonomy and prestige of laboratories, it would perhaps be more helpful to consider—in a less essentializing perspective—whether some laboratories in some circumstances were not so much culturally dominant institutions as interstitial entities, drawing their value and meaning from the people and institutions with which they were juxtaposed, opposed, and integrally networked. (Gooday 2008, 786)

3 Robert Kohler and Henning Schmidgen have pointed out this research gap (Kohler 2008, 762; Schmidgen 2011, 4).

4 "A building set apart for conducting practical investigations in natural science, originally and especially in chemistry, and for the elaboration or manufacture of chemical, medicinal and like products" (*Oxford English Dictionary* 1989, quoted in Gooday 2008, 783).

5 See on the connection between the workshop and the laboratory Fox and Guagnini 1998, on that between the laboratory and the museum, Forgan 1994. Anne Secord has investigated the activities of English artisan botanists in the early nineteenth century in the United Kingdom and thereby pointed out that the role of the public house ("pub") as a locus of scientific discourse and research has hitherto been overlooked (Secord 1994).

Thus, instead of endeavouring to define the laboratory as an autonomous architectural type, research into the history of science came to view it primarily as a hybrid moulded by its social environment and in interaction with other social settings.

There is no doubt that laboratories were of an essentially hybrid nature before the so-called laboratory revolution took place in the latter third of the nineteenth century. A single architectural type does not appear to have existed until laboratories became institutionalised—and thereby increasingly standardised—at European universities. The historian of science Sophie Forgan accordingly noted that: "Until the nineteenth century it is hard to say that any general form existed for a laboratory, although the ideal 'House of Chemistry' featured early in utopian dreams, and the alchemist's chamber was a favoured subject for painters" (Forgan 1989, 409). Laboratories could be found in the most diverse locations, rooms or buildings, and the social standing and hence credibility of these carried over to the experiment and the person conducting it (Kohler 2002, 8). One of the first to demonstrate how a certain understanding of science is closely bound up with the architecture of the laboratory was Owen Hannaway. In a much-cited article from 1989, he compared Tycho Brahe's elite castle laboratory, built around 1600 on the Danish island of Ven, with Andreas Libavius's blueprint, published only a few years later, for a *domus chemiae*: a humanist Utopian combination of chemical laboratory and scientist's residence, which was to be run as an open civic forum (Hannaway 1986) (**Figs. 2.3** and **2.4**). In both cases, according to Hannaway, it was primarily the architectural design that gave expression to a certain ideal of scientific practice. The defensive architecture of Brahe's laboratory spoke of secrecy and exclusion while Libavius's design sought to bring natural sciences research out of the cellar to the ground floor, from the island to the city centre, and hence to make it accessible (at least in theory) to the general public.

For empiricists in seventeenth-century England, as Steven Shapin has shown, the credibility of an experiment depended primarily on the credibility of the person conducting it, which in turn depended on his or her social status (Shapin 1988). Since prestige attached to the origin of any experiments or lines of enquiry,

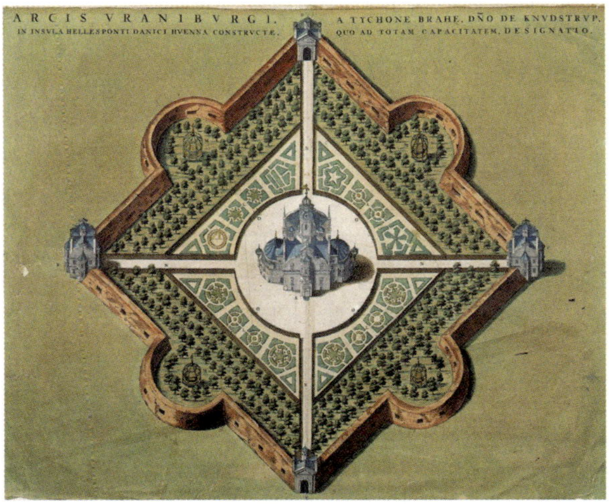

Fig. 2.3:
The Danish Astronomer Tycho Brahe's Palace-Observatory "Uraniborg" on the Island of Ven in Öresund, from: Tycho Brahe, *Astronomiae Instauratae Mechanica*, 1598.

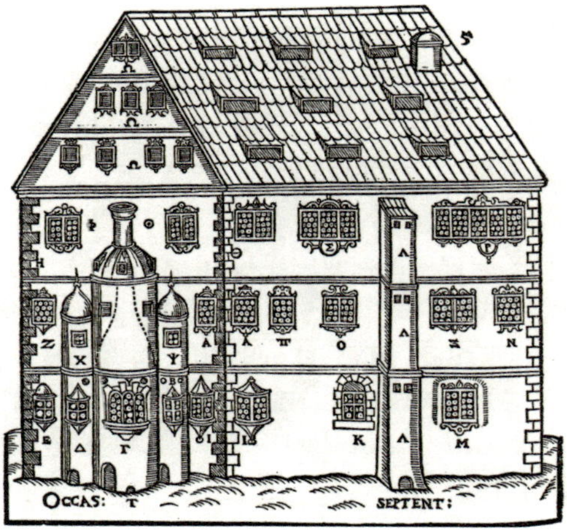

Fig. 2.4:
Andreas Libavius's Idealised Design for his "House of Chemistry", from: Andreas Libavius, *Alchemia*, 1606.

these were pursued in an appropriate social setting—most often in a reputedly trustworthy gentleman's private residence—so as to guarantee their legitimacy (idem). Both Hannaway's analysis of Tycho Brahe's palace observatory and Shapin's examples demonstrate that the experimenters appropriated the respective architectural and social setting as symbolic capital, as a means of better situating their scientific practice in the prevailing social hierarchy. However, in contrast to these isolated laboratories founded at the personal initiative of individual scientists, the first ever "state-funded laboratory system" that began to take shape as of 1870 was, according to Philipp Felsch, "soon to become the linchpin of the modern natural sciences and from its seat in Imperial Germany greatly influence developments in France, England, Italy, the USA and Russia" (Felsch 2005, 28).

The launch of numerous new universities and the growing differentiation between and within each of the natural sciences made empirical instruction increasingly vital (See Chapter 7). Hence, the construction of laboratories at universities and other research institutions accelerated in the 1870s. "Like in the 1960s," the historian of chemistry W. H. Brock remarks in this regard, "the 1870s were a golden age for new academic buildings, not merely in Great Britain, but also in Europe generally" (Brock 1989, 155). General knowledge of laboratory buildings also spread beyond national borders in this period, and laboratories became a favoured destination of scientific research trips. In 1870, the French chemist Adolphe Wurtz presented an exemplary study based on several such laboratory tours, in which he analysed German laboratories and made recommendations for comparable new buildings in France (Wurtz 1870). The technical design and equipment of laboratories were expected to meet international benchmarks at the latest by the early twentieth century. This is especially evident from the detailed lab analyses carried out from 1907 to 1933 by the American physiologist Francis Benedict Gano, who endeavoured to incorporate experimental designs of laboratories all over the world in the construction of his own Laboratory of Physiology (See Chapter 3).

In the late nineteenth century, the first manuals—for chemists as well as for architects—explained how best to equip the increasingly complex laboratories and thereby helped establish international standards in laboratory design. In 1887, the Scottish architect Edward Cookworthy Robins (1830–1918) published his research on technical school and college buildings (Robins 1887), which was to become a standard work of reference on laboratory architecture, as the historian of science William H. Brock has emphasised:

> Robins [...] reflected personally and deeply on the ideal environment for technical education. His book on technical school design continued to be cited by architects until the 1920s; only then was it replaced as the Bible of laboratory design by the works of Russell, Clay and Munby.
> (Brock 1989, 156)[6]

Thus, references to laboratory construction in recent times as generic architecture (Venturi 1999, 388) are a sign not of innovation in the 1980s but of a process that began in the 1880s (See Chapter 5).

There is every indication that the laboratory in the course of its institutionalisation was progressively elaborated as a unique architectural type and then reproduced worldwide. In Robert Kohler's view, this growing trend to "placelessness" marked a striking contrast to the social locus of the pre-institutional laboratory: "since the industrial and social revolutions of the nineteenth century, we have lived in a culture that values universality over locality, and in such a culture placelessness is a reason to trust" (Kohler 2002, 8). Yet this is disputed in a number of studies on the history of laboratories. Rather than viewing the laboratory as a universal, essentially "placeless" building type, historical studies focus instead on those interfaces which call into question the allegedly clear boundaries of the modern laboratory and pin down specific cultural and architectural markers in both its development

6 The reference here is to Russell 1903, Clay 1902 and Munby 1929. For a further early example of the standardisation of teaching laboratories, see the entry "Physikalische Institute" (Institutes of Physics) in *Handbuch der Architektur* (Junk 1888).

and its design. Historians of science have investigated the laboratory of the nineteenth and twentieth centuries and heterogeneous places of experimentation of the seventeenth and eighteenth centuries above all in terms of their permeability vis-à-vis other institutions and production sites. The museum and the factory are two such interfaces, examined more closely below.

Sophie Forgan has demonstrated the symbiotic evolution since the eighteenth century of the natural sciences research laboratory and the museum (Forgan 1994). In the early nineteenth century, laboratories were still commonly located in museums, often in their basements or in ancillary rooms to the rear. Museums were a primary arena for natural science experiments, because a clear distinction between the use and the exhibition of contemporary apparatus had not (yet) been made. Forgan's analysis of architectural plans has shown that the laboratory was permitted more space as of the mid-nineteenth century and thereupon managed to emancipate itself architecturally from the museum. Yet research and exhibition practices remain closely intertwined to this day, as Anke te Heesen and Margarete Vöhringer have shown in their recently published collection of essays *Wissenschaft im Museum—Ausstellung im Labor* (Te Heesen and Vöhringer 2014). The various articles illustrate how museological presentation modes and exhibition practices still pervade scientific laboratories and serve moreover to organise scientific work along aesthetic and practical lines. Despite the ever greater architectural autonomy attained by labs in the course of their institutionalisation, they and likewise museums must still be understood as hybrid spatial types—or so the essays persuasively maintain.

A second interface or analogy recurrently examined by historians of science is that of the laboratory and the factory. Unlike many privately funded laboratories, the ones newly launched at universities in the late nineteenth century were subject to economic constraints. Scientific research procedures were therefore standardised, mechanised and rationalised to the point where they came to resemble assembly-line procedures in certain respects (Schmidgen 2011). Some scientists saw this development with misgivings—yet it also inspired utopian visions. As early as 1892, the English biologist Thomas H. Huxley described the modern university and above all its laboratories as a "factory for new knowledge" (quoted in Fye 1986, 920). Twenty years later, in 1912, Walter Vaughan, dissatisfied with his own physics lab at McGill University in Montreal, not only spoke metaphorically of his ideal lab of the future as a factory building but also based his blueprint on the latter:

> If a University were started with a sufficient area of land I should very much like to see the experiment tried of constructing its buildings somewhat on a factory plan: i.e. having all laboratories one storey (or possibly two storey) buildings in a number of sections with northern skylights. This would not only result in a greater economy of working and supervision, this would allow for expansion at a minimum of cost without detriment to the existing structure. The great fault in the construction of modern University buildings appears to me to be that they are erected as finite structures, so that they cannot be conveniently or economically extended or altered.
> (Quoted in Forgan 1989, 433)

The ideals of flexibility and efficiency associated with the factory building were a radical departure from the image of German laboratories conjured in 1868 by Louis Pasteur as humanist "temple[s] to the future" (quoted in Nuffield Foundation et al. 1961, 7).[7] Until well into the twentieth century, the factory not only served as a rhetorical figure and blank canvas but also decisively influenced the actual design of laboratory buildings (**Fig. 2.5**). Peter Galison and Caroline Jones have shown that the development of "Big Science" labs followed innovations in factory architecture (Galison and Jones 1999, 498). American companies that designed factories prior to and during the Second World War devoted themselves from the 1940s to the 1960s to the development of gigantic

7 In an article published in 1868 in the journal *Moniteur*, Pasteur deplored the slow progress of French laboratories and thereby singled out German laboratories as a commendable model: "I implore you, take some interest in those sacred dwellings meaningly described as laboratories. Ask that they be multiplied and completed. They are the temples of the future, of riches and of comfort. There humanity grows greater, better, stronger" (quoted in Nuffield Foundation et. al. 1961, 7).

laboratory complexes (See Chapter 5). While, until 1941, the MIT Radiation Laboratory was a rather small, intimate and quasi-academic laboratory, under the pressure of urgent wartime research it metamorphosed into "a multibillion-dollar, military-academic-industrial complex" (idem, 504). Flexibility (thanks to modular

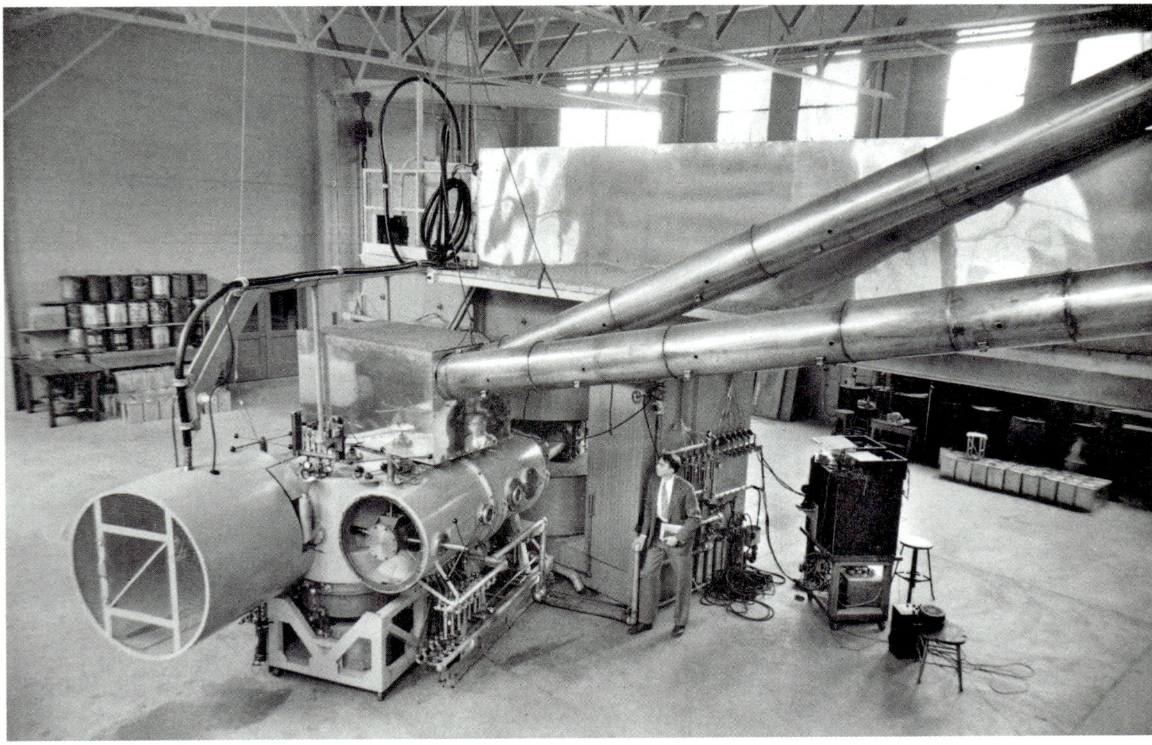

Fig. 2.5:
View of the 60-inch cyclotron that went into operation in 1939, Lawrence Berkeley National Laboratory, Berkeley, California, 1931-39. Photo © 2010 The Regents of the University of California, Lawrence Berkeley National Laboratory.

walls) was as vital a feature of large laboratory sites as of factory halls. The modular design of lab units and the underground installation of utilities (electricity, water, etc.) served both to rationalise working procedures and to make them more easily adaptable to changing circumstances. Henry Smyth, a physicist at Princeton at the time, emphasised in 1945 that: "The laboratory should be essentially of factory-type construction, capable of expansion and alteration. Partitions should be non-structural" (quoted in Galison and Jones 1999, 505). As this close and enduring link with the factory concept shows, the twentieth-century laboratory cannot be defined exclusively in terms of its generic placelessness but derives its credibility also from its interrelation with other places of the Modern era.

A historical survey of laboratory architecture—either of borrowed locations such as the nobleman's castle, the civic forum and the private residence, or in correlation with the museum or factory—attests the existence of extremely disparate laboratory types and thus refutes the essentialist view upheld in the early days of laboratory studies. Graeme Gooday comes to the conclusion that:

> Even if one were to entertain the possibility that there was some special and distinctive rigor associated with laboratory experiments, there has surely been too much diversity of form for us to allow even a single formulaic view of the nature of the laboratory.
> (Gooday 2008, 788)

Inquiry into the diversity of and "structural variations" in laboratory architecture, such as Karin Knorr Cetina has called for, is particularly relevant today (Knorr Cetina 2002a, XVIII). Although flexibility and efficiency, two criteria borrowed from the factory floor, remain decisive for the design of modern research buildings, discourse on the latter has radically changed since the 1940s and 1950s. Interdisciplinary communication as well as the need to provide research scientists with a pleasant work environment are the new paradigms of lab design, and are more strongly oriented to the rhetoric of New Economy architecture than to traditional factory structures (See Chapter 4).

Yet a parallel trend has been noted too in recent years—namely the gradual disappearance of the laboratory as an actual physical space—and this once again raises the question of the locus of experimentation. It seems as if the laboratory today is "in a process of dissolution and dispersion", as Henning Schmidgen highlights:

> Above all the large research centres for particle physics, such as the Fermilab near Chicago or the CERN in Geneva Canton, and the scientific enterprises that address current issues in biology, the *Human Genome Project*, for example, have contributed to a network-type expansion of laboratory practice, which is now very far removed from the traditional image of the *tabletop experiment* conducted in a closed and isolated space. (Schmidgen 2011, 1)

Peter Galison and Caroline Jones likewise draw on the CERN as an example of how major research projects increasingly unfold beyond the architectural boundaries of any single laboratory: "if the control rooms are multiple and the data are in play across the world, it is no longer clear *where* the experiment can be said to be. The experiment cannot be localized for the circus tent spans the globe" (Galison and Jones 1999, 529).

Yet a contrary trend in contemporary lab architecture is likely to halt the latter's anticipated dissolution. Natural science research institutions attuned to the *branding* potential of distinctive lab design now seek to reap the benefits of striking symbols of innovation that draw international research scientists like a magnet (See Chapter 5). The laboratory—often consciously staged as a "building set apart"—is increasingly a factor that shapes both research and the research environment. The physical locus of experimentation is currently not only "in a process of dissolution and dispersion" but also—as the essays in this present volume make clear—taking centre stage as an important factor in its own right. Whether the dissolution and simultaneous high-profile construction of laboratories are mutually exclusive trends remains to be seen. Perhaps, one might conjecture, it is precisely the global reach of networked data, the virtual expansion of research beyond national borders, that makes a consolidation of the physical location now seem more necessary to scientific institutions than ever before.

The answers formulated in response to the question "What is a laboratory?" could not be more diverse and contradictory. While sociologists and anthropologists engaged in laboratory studies have endeavoured to define the laboratory as a "place set apart", historians of science have sought to identify those interfaces at which the laboratory occurs as a hybrid and hence defies unambiguous definition. Knorr Cetina's call to make the analysis of both historical and contemporary varieties of the laboratory a complementary sub-discipline of laboratory studies therefore appears at first glance to be a paradoxical enterprise. It does, however, make patently clear that it is impossible to answer the question "What is a laboratory?" from a mono-disciplinary perspective.

REFERENCES

Anderson, Robert G. W. 2013. "Chemistry Laboratories, and How They Might Be Studied." *Studies in History and Philosophy of Science Part A* 44 (4): 669-75.

Brock, William H. 1989. "Building England's First Technical College: The Laboratories of Finsbury Technical College, 1878-1926." In *The Development of the Laboratory. Essays on the Place of Experiment in Industrial Civilization*, edited by Frank A. J. L. James, 155-70. London: Macmillan Press.

Brockhaus Enzyklopädie. 2006. 21st ed., Leipzig: F. A. Brockhaus.

Busch, Lawrence, Jeffrey Burkhardt, and William Lacy. 1991. *Plants, Power, and Profit: Economic and Ethical Consequences for the New Biotechnologies*. Oxford: Basil Blackwell.

Clay, Felix. 1902. *Modern School Buildings: Elementary and Secondary*. 1st ed., London: B. T. Badsford.

Felsch, Philipp. 2005. "Das Laboratorium." In *Orte der Moderne: Erfahrungswelten des 19. und 20. Jahrhunderts*, edited by Alexa Geisthövel and Habbo Knoch, 27-36, 1st ed., Frankfurt / Main: Campus Verlag.

Forgan, Sophie. 1989. "The Architecture of Science and the Idea of a University." *Studies in History and Philosophy of Science* 20 (4): 405-34.

Forgan, Sophie. 1994. "The Architecture of Display: Museums, Universities and Objects in Nineteenth-Century Britain." *History of Science* 32 (2), 139-162.

Foucault, Michel. 1992. "Andere Räume." In *Aisthesis: Wahrnehmung heute oder Perspektiven einer anderen Ästhetik*, edited by Karlheinz Barck, Peter Gente, Heidi Paris, and Stefan Richter, 34-46. Leipzig: Reclam.

Fox, Robert and Anna Guagnini. 1998. "Laboratories, Workshops, and Sites: Concepts and Practices of Research in Industrial Europe, 1800-1914." *Historical Studies in the Physical and Biological Sciences* 29 (1): 55-139.

Frame, J. Davidson, Francis Narin, and Mark P. Carpenter. 1977. "The Distribution of World Science." *Social Studies of Science* 7 (4): 501-16.

Fye, W. Bruce. 1986. "Carl Ludwig and the Leipzig Physiological Institute: 'A Factory of New Knowledge'." *Circulation* 74 (5): 920-28.

Galison, Peter and Caroline A. Jones. 1999. "Factory, Laboratory, Studio: Dispersing Sites of Production." In *The Architecture of Science*, edited by Peter Galison and Emily Thompson, 497-540. Cambridge, MA: MIT Press.

Gilpin, Robert and Christopher Wright, eds. 1964. *Scientists and National Policy Making*. New York: Columbia University Press.

Gieryn, Thomas F. 2008. "Laboratory Design for Post-Fordist Science." *Isis* 99 (4): 796-802.
Gooday, Graeme. 2008. "Placing or Replacing the Laboratory in the History of Science?" *Isis* 99 (4): 783-95.
Hannaway, Owen. 1986. "Laboratory Design and the Aim of Science: Andreas Libavius versus Tycho Brahe." *Isis* 77 (4): 585-610.
Hendersen, Kathryn. 1999. *On Line and on Paper.* Cambridge, MA: MIT Press.
Junk, Carl. 1888. "Physikalische Institute." In *Handbuch der Architektur*, edited by Josef Durm, Hermann Ende, Eduard Schmitt, and Heinrich Wagner, vol. 4, 101-58. Darmstadt: Verlag von Arnold Bergsträsser.
Knorr Cetina, Karin. 1981. *The Manufacture of Knowledge.* Oxford: Pergamon Press.
Knorr Cetina, Karin. 1995: "Laboratory Studies. The Cultural Approach to the Study of Science." In *Handbook of Science and Technology Studies*, edited by Sheila Jasanoff, Gerald E. Markle, James C. Peterson, and Trevor Pinch, 140-66. London: Sage Publications.
Knorr Cetina, Karin. 2002a: "Laborstudien. Vorwort zur Neuauflage." In Karin Knorr Cetina. *Die Fabrikation von Erkenntnis: Zur Anthropologie der Naturwissenschaft.* XI-XXIV. 2nd ed., Frankfurt / Main: Suhrkamp.
Knorr Cetina, Karin. 2002b: *Wissenskulturen: Ein Vergleich naturwissenschaftlicher Wissensformen.* 1st ed., Frankfurt / Main: Suhrkamp.
Kohler, Robert E. 2002. *Landscapes and Labscapes.* Chicago: University of Chicago Press.
Kohler, Robert E. 2008. "Lab History: Reflections." *Isis* 99 (4): 761-68.
Korach, Maurice. 1964. "The Science of Industry." In *The Science of Science*, edited by Maurice Goldsmith and Alan Mackay, 179-94. London: Souvenir Press.
Latour, Bruno and Steve Woolgar. 1986. *Laboratory Life: The Construction of Scientific Facts.* 2nd ed., Princeton, NJ: Princeton University Press.
Latour, Bruno. 1988. "Extrait d'une présentation du terrain." In *La vie de laboratoire. La production des faits scientifiques*, edited by Bruno Latour and Steve Woolgar, 10-15. Paris: Editions La Découverte.
Latour, Bruno. 2004. "Von 'Tatsachen' zu 'Sachverhalten': Wie sollen die neuen kollektiven Experimente protokolliert werden?" In *Kultur im Experiment*, edited by Henning Schmidgen, Peter Geimer and Sven Dierig, 17-36. Berlin: Kadmos.
Leslie, Thomas. 2005. *Louis I. Kahn: Building Art, Building Science.* New York: George Braziller.
Lynch, Michael. 1985. *Art and Artifact in Laboratory Science: A Study of Shop Work and Shop Talk in a Research Laboratory.* London: Routledge.
Merleau-Ponty, Maurice. 1962. *Phenomenology of Perception.* Translated by Colin Smith, London: Routledge.
Munby, Alan E. 1929. *School Laboratory Fittings.* London: G. Bell & Sons.
Nuffield Foundation / Division for Architectural Studies, Leslie Farrer-Brown and Richard L. Davies, eds. 1961. *The Design of Research Laboratories.* London: Oxford University Press.
Oxford English Dictionary. 1989. Edited by John A. Simpson and Edmund Wiener, vol. 8, 2nd ed., Oxford: Oxford University Press.
Oxford English Dictionary. 2014. <http://www.oed.com> (accessed 20 February 2015).
Robins, Edward C. 1887. *Technical School and College Building.* London: Whittaker and Co.
Russell, Thomas H. 1903. *The Planning and Fitting-up of Chemical and Physical Laboratories.* London: B. T. Batsford.
Schmidgen, Henning. 2011. "Labor." *Europäische Geschichte Online.* <http://www.ieg-ego.eu/de/threads/crossroads/wissensraeume/henning-schmidgen-labor> (accessed 10 February 2015).
Secord, Anne. 1994. "Science in the Pub: Artisan Botanists in Early Nineteenth-Century Lancashire." *History of Science* 32 (3): 269-315.
Shapin, Steven. 1988. "The House of Experiment in Seventeenth-Century England." *Isis* 79 (3): 373-404.

Smith, Robert W. and Joseph N. Tatarewicz. 1985. "Replacing a Technology: The Large Space Telescope and CCDs." *Proceedings of the IEEE*, 73 (7): 1221-35.

Te Heesen, Anke and Margarete Vöhringer, eds. 2014. *Wissenschaft im Museum - Ausstellung im Labor*. Berlin: Kadmos.

Traweek, Sharon. 1988. *Beamtimes and Lifetimes: The World of High Energy Physicists*. Cambridge, MA: Harvard University Press.

Venturi, Robert. 1999. "Thoughts on the Architecture of the Scientific Workplace: Community, Change and Continuity." In *The Architecture of Science*, edited by Peter Galison and Emily Thompson, 385-98. Cambridge, MA: MIT Press.

Wurtz, Adolphe. 1870. *Les Hautes Études Pratiques dans les Universités Allemandes*. Paris: Imprimerie Impériale.

THE IMAGE OF THE LABORATORY

“People travel, get ideas, bring them back home and implement them. This is true for researchers as well as for architects. It leads to the homogenisation of labs and lab buildings.”

Antoine Kahn, Department of Electrical Engineering, Princeton University, USA

3

SCIENCE IN PICTURES: A HISTORICAL PERSPECTIVE

Verena Straub

The chemistry lab has long since outgrown its image as the sterile production site of scientific research and even serves at times as a stage for successful TV dramas, most recently for the American cult series *Breaking Bad* (2008–13). Here, a chemistry teacher in cahoots with his drug-dealing student becomes the ultimate anti-hero in his quest to produce the purest crystal meth in laboratories as improvised as they are risky. Although in reality crystal meth is transparent, in *Breaking Bad* it is shown to be luminescent blue, a colour that has since become the drug's popular trademark. In April 2010, in the US state of Utah, narcotics agents making a raid came across an unusual blue crystal meth: food dye had been used to imitate the *Breaking Bad* effect (Drösser 2013). Images of laboratories have evidently influenced the drug scene that originally inspired them and can shape not only our imaginations but also reality itself.

Yet our image of the laboratory stems not from Hollywood alone, but is rooted also in a centuries-old iconographic tradition that remains predominant in our cultural imagination to this day (Schmidgen 2011). All the more surprising, then, that the iconography of the modern laboratory remains a thoroughly neglected topic of research. True, one finds numerous studies of the so-called alchemist portraits, a genre popular in the seventeenth century, in particular among Flemish painters; yet very little has been written about the representation of the laboratory in the wake of its institutionalisation and the attendant differentiation of scientific disciplines in the nineteenth century. Around 1900, however, shortly after the natural science lab had become an established feature of European universities, an abundance of photographs was taken to document lab interiors. Viewed as a whole, these photographs attest the co-existence of various and, sometimes, disparate iconographies of the modern laboratory.

PORTRAITS IN A LABORATORY SETTING

Photographic images of laboratories quickly proved to have a powerful impact. A photograph published in the popular science book *The Romance of Modern Electricity* (1906) by Scottish author and lay scientist Charles R. Gibson shows him in a "corner of [his] laboratory" (**Fig. 3.1**). According to the historian of science Graeme Gooday, the corner in question is located in a "domestic room, as is readily indicated by the wallpaper on the right" (Gooday 2008, 792). In his article "Placing or Replacing the Laboratory in the History of Science?" Gooday calls for a broader definition of the laboratory, one not limited to institutional labs alone, but cognizant also of the fact that experiments were carried out in kitchens, workshops, hospitals—or, as in the case of the lay scientist Gibson, even in private homes (See Chapter 2). Gibson's self-portrayal is of interest, not only in respect of the more inclusive approach to a history of laboratories propounded by Gooday, but also, and more emphatically, on account of the lab iconography it depicts. Gibson, who gained renown for popular science publications on electromagnetism, strikes a pose in this portrait, in an evident endeavour to convey professional standing. As Gooday remarks, Gibson uses the portrait strategically to stage himself as a man of science to be reckoned with: "Gibson ostentatiously adopts the authoritative role of the expert scientist by donning a white laboratory coat and positions himself leaning over the iconic instrument of laboratory epistemology: the microscope" (idem). Gibson indeed appears here to have consciously referenced the popular portrait tradition of his day, namely to prominently feature a white coat, a microscope and test tubes, which around 1900 ranked among the most iconic attributes of natural scientists, be they chemists, physicists or physiologists.

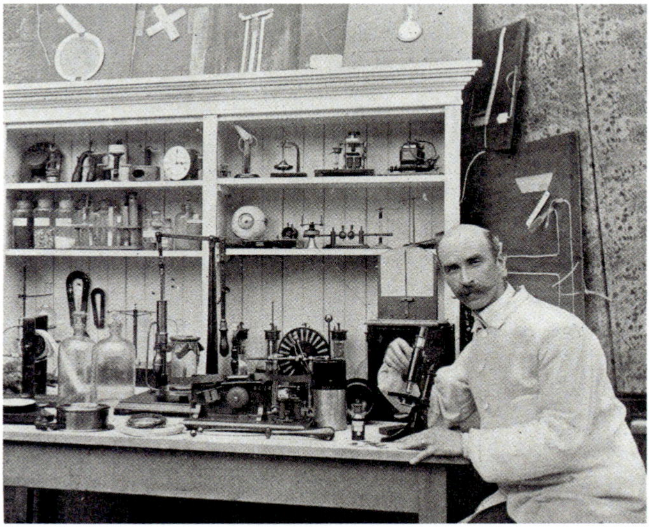

Fig. 3.1:
Popular Scottish Science Writer Charles Gibson "in a corner of [his] laboratory", from: Gibson 1906: 326.

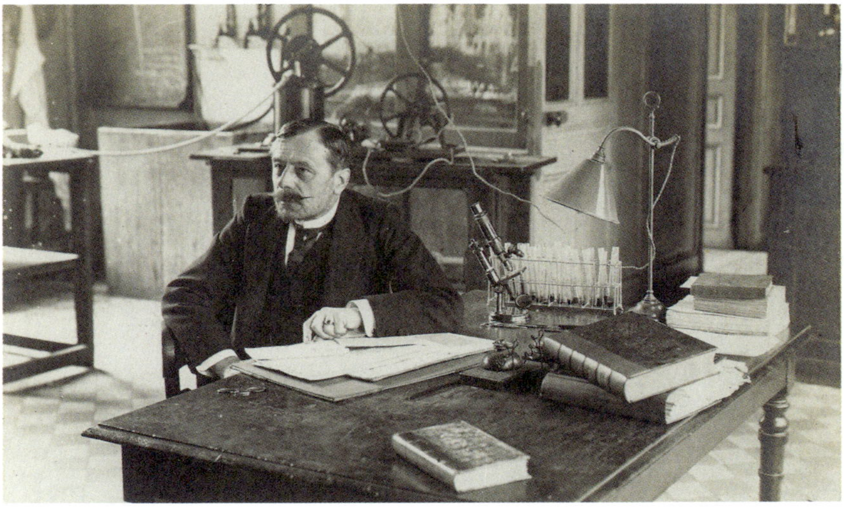

Fig. 3.2:
"Prof. Dastre, Director of the Laboratory of Physiology, in the Sorbonne", from: Benedict 1907, 1: 294. Photo © Harvard University Archives.

A comparable iconography is apparent, for instance, in a photograph from 1907 portraying Professor Dastre, then Director of the Laboratory of Physiology at the Sorbonne. It shows him in a similar pose to Gibson, at a desk on which open books, a microscope and test tubes are pretentiously arranged like props, as symbols of the laboratory's experimental projects (**Fig. 3.2**).[1] In Gibson's photograph, this staged setting is furthermore highlighted by his white coat: hardly justifiable on hygienic grounds as it is so far removed from a lab bench and with neither utilities nor even chemicals in view, it evidently serves an iconic purpose. The numerous mechanical apparatus, models and empty glass flasks displayed on the table and cabinet likewise appear to be exhibits laid out expressly for the photo. Gibson's staged self-portrait not only draws on the iconographic conventions of the scientist portraits of his day but also deploys them consciously as a strategy of legitimation. Instead of simply reflecting the reality of his work on electromagnetism, Gibson aims to purvey a familiar image of the laboratory scientist to his readers and thus persuade them that his research is no less professional than that of prestigious institutions.

Representations of laboratories, as this example unequivocally demonstrates, are always also to be read as imaginary settings that strive to conform to a certain ideal. As fictions, images of laboratories give expression to a specific notion of science and therefore cannot be regarded unchallenged as records of authentic laboratory experience. As early as 1975, Christopher Hill drew attention to the risk of interpreting portrayals of laboratories as purely technical illustrations. Etchings, drawings and paintings of the lab of the Early Modern era, a motif which became popular in the seventeenth century among Flemish masters such as David Teniers the Younger or Thomas Wijk, were regarded by historians of chemistry until well into the 1960s as credible sources from which to glean information about the technical equipment and working methods employed in historical laboratories. Hill was one of the first to contend that portrayals of alchemistic experimental settings

1 For a further laboratory portrait with similar iconography, see Benedict 1913, 3: 302.

are iconographic formulae, to be read not as sources of historical knowledge but in the light of a tradition of genre painting (Hill 1975).

Nonetheless, the impact of such formulaic alchemist portraits can still be traced in the lab photography of the early twentieth century, where the genre was perpetuated through the figure of the solitary genius. This iconographic tradition is referenced in a particularly striking manner in the famous photograph from 1899 of Nikola Tesla in his laboratory in Colorado Springs (**Fig. 3.3**). Here, the camera captures not only a spectacular show of dazzling sparks emitted by the gigantic Tesla coil's discharge of electricity, but also, next to it, the scientist seated in a chair, engrossed in a book. This image—one of a series shot by the photographer Dickenson Alley from 17 to 31 December 1899 on behalf of *Century Magazine*—was the result of a double exposure. Alley first photographed the electrical sparks in an otherwise unlit laboratory then exposed the same photographic plate a second time, now with the coil inactive and Tesla seated on his chair.[2] Tesla himself wrote of a similar shot from the same series:

> To give an idea of the magnitude of the discharge the experimenter is sitting slightly behind the 'extra coil'. I did not like this idea but some people find such photographs interesting. Of course, the discharge was not playing when the experimenter was photographed, as might be imagined.
> (Tesla 2007, 367)

Peter Galison and Caroline Jones see in this staged publicity shot a return to the Romantic notion of the solitary genius, a topos previously present in the idealised image of the alchemist labouring alone over experiments (Galison and Jones 1999, 508–9). The

2 That this photo was the result of a double exposure was emphasised also by Tesla's biographer, Marc Seifer: "By using multiple exposures, Alley would create what is perhaps Tesla's most famous photograph: that of the inventor sitting calmly reading a book, dwarfed by myriad tongues of explosive lightning. (This picture is a multiple exposure. Tesla, of course, was not sitting there at the time the oscillator was fired up; the electricity would have killed him.)" (Seifer 1998, 234).

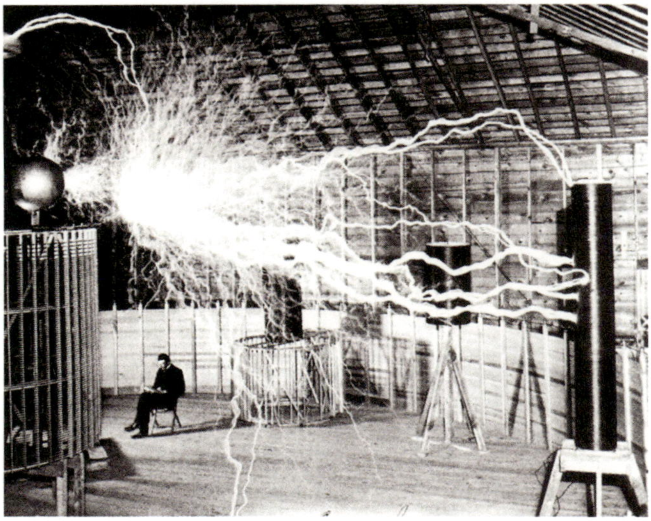

Fig. 3.3:
Dickenson Alley, *Nikola Tesla in his Laboratory in Colorado Springs*, from: Tesla 2007: Plate XII.

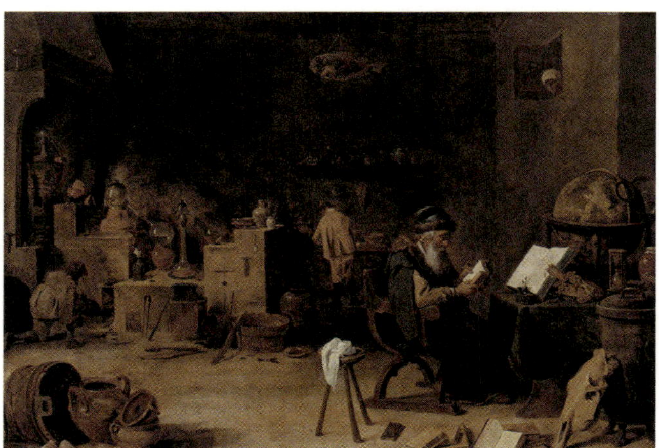

Fig. 3.4:
David Teniers II. (1610-1690), *The Alchymist*, oil on wood, 51×71 cm. Photo © Herzog Anton Ulrich Museum Braunschweig.

double exposure does indeed seem to be a modern elaboration of the motif identified by Hill: the *alchemist-philosopher* portrayed among his books, globes and technical apparatus (Hill 1975, 109), especially when compared with David Teniers the Younger's painting *The Alchymist*, from circa 1645 (**Fig. 3.4)**. There too, the scientist is shown in a rapt pose, engrossed in his books, with a chemical experiment bubbling away in the background as a symbol of his creative genius. Yet in addition to its patently conscious staging of this centuries-old motif, the photograph of Tesla conjures a second modern myth, namely that of the laboratory as factory, which has thrived in the natural sciences context since the late nineteenth century (See Chapter 2). In contrast to the modest room of Teniers's alchemist, the hangar-like laboratory with its gigantic oscillator dwarfs Tesla and makes him seem a humble figure. Two disparate lab motifs—the lonely alchemist and the factory—are fused in this one image and thus presage the laboratory's future development, which Galison and Jones have described as a shift from the "epoch of humanistic individuality" to the centralised structures of major industrial laboratories after 1945 (Galison and Jones 1999, 498).

The photographs of Tesla modified the image of the alchemist, who had traditionally been portrayed among chemical experiments, bubbling gases and variously shaped glass flasks and vials. Now, for the first time, the spotlight was turned on electricity instead. Repeatedly, Tesla was portrayed as a quasi-alchemical electricity-conjurer, a magician unable to be harmed even by one million volts and mysteriously able to switch on a light bulb with nothing more than a profoundly contemplative gaze. This image of the eccentric genius able to spectacularly direct electrical currents ultimately inspired filmmakers' imaginations too, as Fritz Lang's *Metropolis* from 1927 most vividly attests. In a famous scene in Rotwang's laboratory, the robotic figure Maria is subjected to an electrical storm that may well have been modelled on the famous double exposure from Tesla's lab in Colorado Springs.

EXPERIMENTATION IN PROGRESS

This impressive modern reinterpretation of the alchemist motif was one among a broad spectrum of laboratory images produced in the wake of the "laboratory revolution" of the late nineteenth century (Schmidgen 2011, 20–23). The rapid worldwide standardisation of laboratory architecture, which began in the 1870s and prompted a search for the *ideal* laboratory, led to a plethora of laboratory images, mostly of a documentary nature (idem, 24–25). At the time, scientists began to visit laboratories all over the world in order to gather information useful to their own architectural projects, information that was increasingly conveyed by visual means. The images resulting from these laboratory trips initially included sketches of architectural plans, such as the chemist Adolphe Wurtz produced for his report on German university laboratories (Wurtz 1870). Perhaps the most outstanding collection of all, however, is the several hundred photographs taken by the American physiologist, chemist and nutritionist Francis Gano Benedict, who travelled to laboratories all over Europe in search of inspiration for the design and equipment of the Laboratory of Physiology in Boston, of which he was appointed Director in 1907. The Laboratory, newly founded by the Carnegie Institution of Washington and soon to gain international renown as the Nutrition Lab, was located in a building directly alongside the Harvard Medical School. In order to keep abreast of advances made in European experimental physiological research, Benedict visited more than fifty laboratories between 1907 and 1933, and meticulously documented their design, apparatus and working procedures. In Benedict's published work, seven volumes in total, photographs accompanied by detailed notes are a fundamental feature. They not only constitute a "photographic panorama of European physiology laboratories" (Neswald 2010, 4) but also are of interest in their own right inasmuch as they comprise a mix of three different types of laboratory image, which respectively attest representative, documentary or aesthetic aspirations.

In parallel to the formulaic scientist iconography exemplified by the portrait of Professor Dastre (**Fig. 3.2**), Benedict increasingly documented the everyday life of laboratories, thereby producing a visual record of experimentation in progress. In the preface to his second volume, he noted that it was only by personally observing the progress of an experiment that one might ever grasp all the detail and methodological nuances involved (Benedict 1910). Although not referring here to his own photographs explicitly, it is clear that Benedict considered photography an adequate means of recording those "important minor details which, though they may not affect the principle of the apparatus, nevertheless play a very important role in the successful conduct of experiments" (idem, 1).

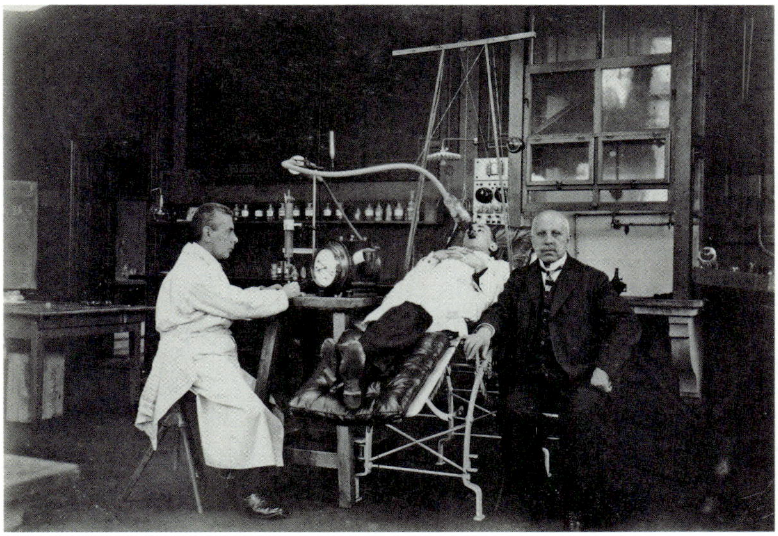

Fig. 3.5:
"Determination of the respiratory exchange. Dr. Abelin, left, and Professor Asher, right, of Berne, Switzerland", from: Benedict 1926, 5: 136. Photo © Harvard University Archives.

He was interested not only in technical apparatus and how these were installed in the laboratory but also, and above all, in the interaction of man and machine. In particular the resistance evident in the respiration experiments documented by Benedict—which measured the test person's respiratory volume with the aid of tubes, pressure chambers and spirometers—brings to mind that which Andrew Pickering has described as the "dance of agency"

(Pickering 1995, 21). Pickering argues that scientific practice must be regarded as a performative process in which not only human beings but also machines and material constitute “an evolving field of human and material agencies reciprocally engaged in a play of resistance and accommodation in which the former seeks to capture the latter” (idem, 23). While the laboratory portrait is an idealising genre that foregrounds the scientist as a sovereign agent and linchpin, Benedict’s quasi-documentary photographs portray the reality of laboratory life in its performative dimension.

Things become particularly interesting when these two types of portrayal are fused, as in a photograph taken at the Physiology Laboratory at Berne University—not by Benedict himself, as the signature shows, but by one of the scientists engaged there, who later sent it to him (**Fig. 3.5**). In the left half of the image, a scientist in a white coat is conducting an experiment on a test person as part of his research into human respiration. To his right, a professor dressed in a suit and tie adopts a distinguished stance, rests his hand on the bench and gazes directly into the camera. This photograph evidently aimed not so much to document the experiment underway but rather to stage it merely as a backdrop to the conventional portrait in a laboratory setting. The experiment here plays a secondary role, as a defining attribute of the illustrious professor in command of the situation.[3]

AESTHETICISATION OF THE LABORATORY

In contrast to the latter, Benedict published an entire series of photographs in which humans are entirely absent and the focus is on individual details of laboratory apparatus. Such photographs

3 A tendency to aestheticisation can be noted likewise in the laboratory illustrations of the German physiologist Emil Du Bois-Reymond, which he included in his scientific publications (*Thierische Electrizität, Vols. 1 and 2*, 1848-49). According to the historian of science Sven Dierig, the self-portraits in which Du Bois-Reymond gave himself the idealised physical anatomy of a classical sculpture must be read as a reference to the classical motif of the youthful, athletic Apollo. In this image, Dierig argues, the classical aesthetic is combined with a contemporary understanding of education and experimental physiology (Dierig 2002, 4).

primarily served operational purposes—for instance, objects were photographed in combination with measuring tapes or yardsticks so as to convey their actual size—yet certain photographs also evince intrinsic aesthetic qualities. One photograph Benedict took in a physiology lab in Florence shows the individual technical components of a respiratory machine laid out on a table as an

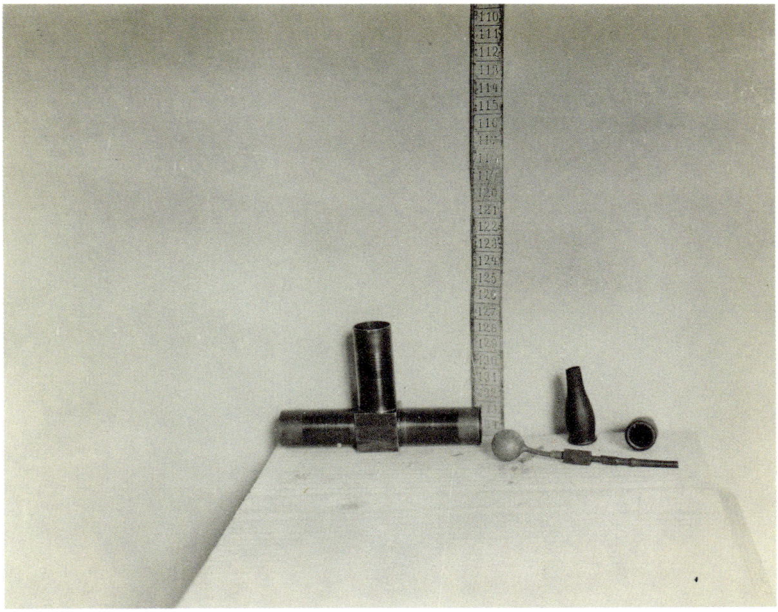

Fig. 3.6:
"Part of the Marchetti valve, the rubber balloon for the Morelli valve, and rubber nosepieces for the Morelli combination mouth- and nosepiece", from: Benedict 1913, 3: 98. Photo © Harvard University Archives.

abstract composition, with a measuring tape running vertically through the image dividing it almost in the golden ratio. This arrangement according to the size of ventilators, nosepieces and rubber balloons lends these banal objects the allure of a carefully arranged still life (**Fig. 3.6**). Other photographs make no pretension to documentation at all, for example, the one Benedict took of the “indistinct details” of the Zuntz gas analysis machine at the Laboratory of Physiology in Berlin (Benedict 1913–16, 98).

Except for diffuse reflections in a glass surface behind which blurred milky glass tubes are only just discernible, the photograph shows nothing that clearly conveys information about the way the machine works. Why would Benedict select such a photograph to illustrate his technical reports when he himself remarked in the caption that it reveals no clear details of note? Benedict did not offer any written explanation. However, the images themselves speak volumes about his great or perhaps even primary interest in the aesthetics of technical apparatus. The protagonist in this image

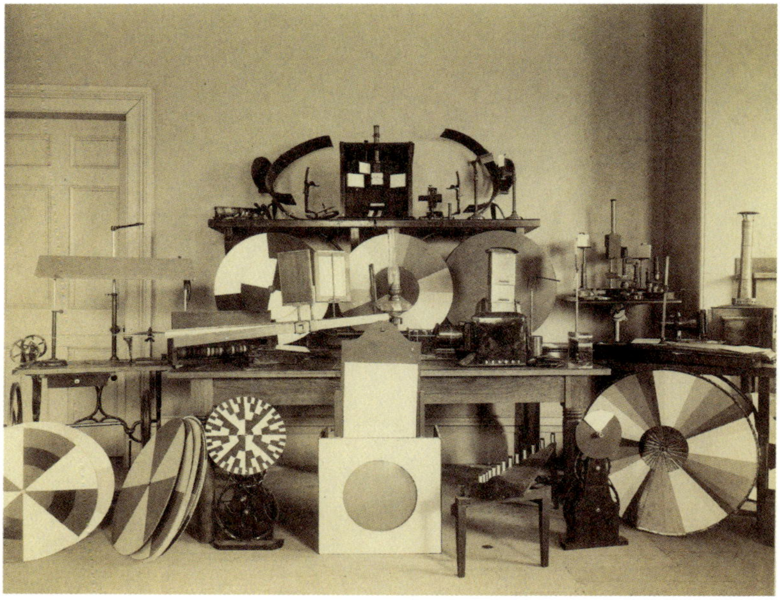

Fig. 3.7:
Harvard Psychological Laboratory in Dane Hall: Instruments for Experiments on Sight, 1892. Photo © Harvard University Archives.

is neither the scientist nor the interaction of man and machine but the apparatus itself, conspicuously showcased in its purely aesthetic dimension.

Henning Schmidgen has noted a similar shift in focus in the laboratory photographs taken by the German psychologist Hugo Münsterberg (Schmidgen 2008b). In 1893, on the occasion of the World Exhibition in Chicago, Münsterberg exhibited images of his new psychology laboratory at the Harvard University Laboratory. All the 240 instruments and objects he had acquired for his new lab were listed in a thirty-five-page catalogue, along with eight photographs of its interior design and layout. These images differ fundamentally from the conventional group portrait believed to be the only surviving visual record of Münsterberg's period of research in Freiburg, Germany. As Henning Schmidgen has noted, three of the Harvard photographs show instead "'psychology in action', i.e. the interaction of persons and things in the process of psychological experimentation" while the other five are "authentic still lifes depicting artfully arranged instruments for specific areas of psychological research (sight, hearing, perception of space, time measurements of mental acts, etc.)" (idem, 6). In one of the latter, instruments of experimental research into human sight are arranged in order of size in the form of a pyramid (**Fig. 3.7**). The two uppermost, semi-circular perimeters serve as a perfect frame for the square apparatus used to diagnose colour blindness, which is at the centre of the composition. Evidently, the point here was not merely to make a documentary record of the newly acquired apparatus but to present them rather as an artful entity without in any way inferring their actual use in scientific practice.

In Schmidgen's view, the iconographic shift from the idealising scientist portrait to a focus both on experiments in progress and on abstract arrangements of apparatus devoid of a human presence above all reflects "a decisive shift in [Münsterberg's] psychological research practices: from a cognitive and/or idealist 'Physiological Psychology' in the sense of Wundt to the pragmatist and/or functional 'Science of Mental Life', as advocated by James and others" (idem, 10). The altered material reality of Münsterberg's laboratory, as depicted in the photographs, indicates changes in Münsterberg's understanding of psychological research practice, Schmidgen argues. A similar thematic correlation cannot be easily traced in Benedict's photographs, however. His seven-volume work on the laboratory evinces not an iconographic evolution, but rather the simultaneous juxtaposition of three iconographic types: the *conventional scientist portrait,* the *experiment in progress* and the *aestheticised still life*. Moreover, these three iconographic types are merged at times within a single image (**Fig. 3.5**).

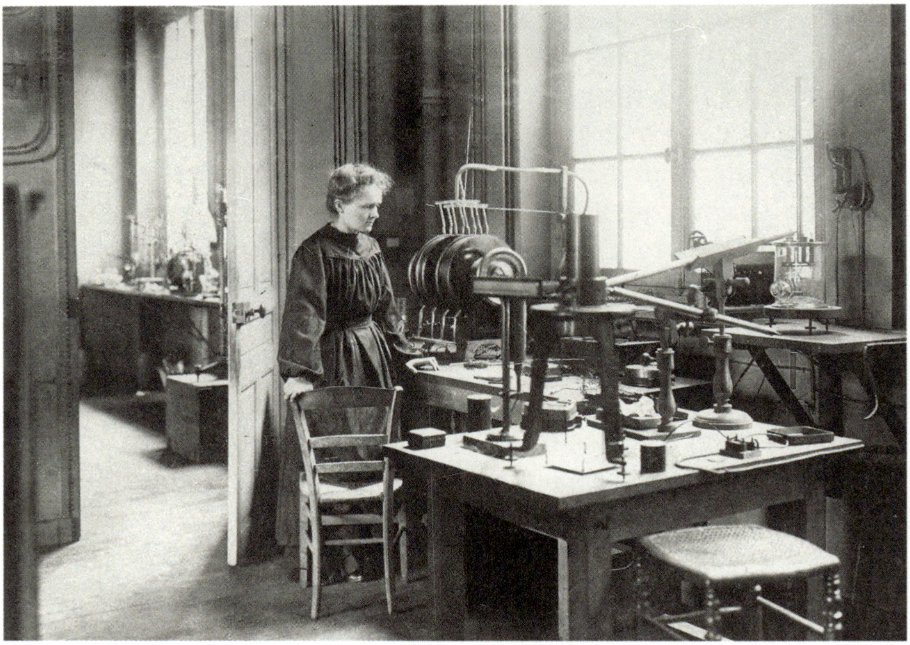

Fig. 3.8:
Marie Curie (1867-1934) in her Laboratory in the Sorbonne, Paris, 1911.
Photo © The LIFE Picture Collection / Getty Images.

That the laboratory in its role as showcase for aesthetic arrangements became an increasingly common motif in photography of the early twentieth century is attested inter alia by the image of Gibson's domestic laboratory discussed at the start of this chapter (**Fig. 3.1**). Here, the wooden cabinet packed full of various apparatus is framed furthermore by flat artefacts—models of circuit boards, by all appearances—which, hung on a wall like artworks, have an aesthetic somehow reminiscent of Constructivist paintings. In similarity to Münsterberg's still lifes, Gibson's photograph does not aspire to be read as a realistic documentation of an experiment in progress. On the contrary, in portrayals such as these, the laboratory serves as a stage for aesthetic encounters that appear in certain cases to border on the mythical, as for instance in the famous photograph of Marie Curie in her laboratory at the Sorbonne, taken in 1911 (**Fig. 3.8**). In striking contrast to the distinguished pose and direct gaze into the camera that was so commonly staged, Curie here appears to be engaged in nothing less than reverent dialogue with the apparatus that she and Pierre Curie had developed for measuring radioactivity. Standing on the laboratory workbench, quasi eye to eye with the scientist, the apparatus appears to be her equal,[4] and under her intense, unerring gaze, it rivets the viewer's attention, as much a protagonist of the photograph as Curie herself.

CONTEMPORARY LABORATORY IMAGES

What this kind of aestheticisation of the laboratory, this emphasis on the sculptural or still life qualities of technical apparatus reveals about scientific practice cannot be conclusively stated. Yet that the aesthetic evinced by laboratory images from around 1900 continues to resonate also in contemporary artworks is beyond any doubt. Large-format art prints by the German photographer Thomas Struth come to mind, in which sections of the particle accelerator BESSY II in Berlin are portrayed as aesthetically imposing structures. Of his fascination with complex physical experimental structures that have evolved over years or even decades Struth said: "I see these scenarios as the result of human activity, as the imprint of a mental process, a group dynamic, represented in sculptural form" (Kuhn 2014). The way in which Struth discusses and portrays laboratory apparatus as aesthetic sculpture is indeed reminiscent of the photographs discussed above. In this regard, Struth's aestheticisation of the laboratory is not a new

4 That Marie Curie does not gaze directly into the camera may also indicate a gender-specific staging convention. Graeme Gooday uses another example to draw attention to differences between portrayals of males and females in the laboratory setting. He compared the portrait of Gibson discussed at the start of this chapter with a picture taken circa 1912, likewise in a home laboratory, of the English mathematician and electrical engineer Hertha Ayrton. "While Gibson gazes directly at the camera," Gooday notes, "the noble widow looks modestly away; and she wears no white coat to claim scientific authority" (Gooday 2008, 793). The extent to which and the reasons why female scientists in laboratory settings were portrayed differently than their male counterparts remains a research subject of great interest, investigation of which with respect to laboratory iconography would doubtless be most instructive.

departure, but stands in an iconographic tradition with roots in the late nineteenth century.

While in the artwork discussed above the staged visual motif is clearly the physical apparatus of experimentation, it remains to be seen which images of the scientist shape our imagination today. The image of the demonic chemist (*Dr Jekyll and Mr Hyde*, dir.

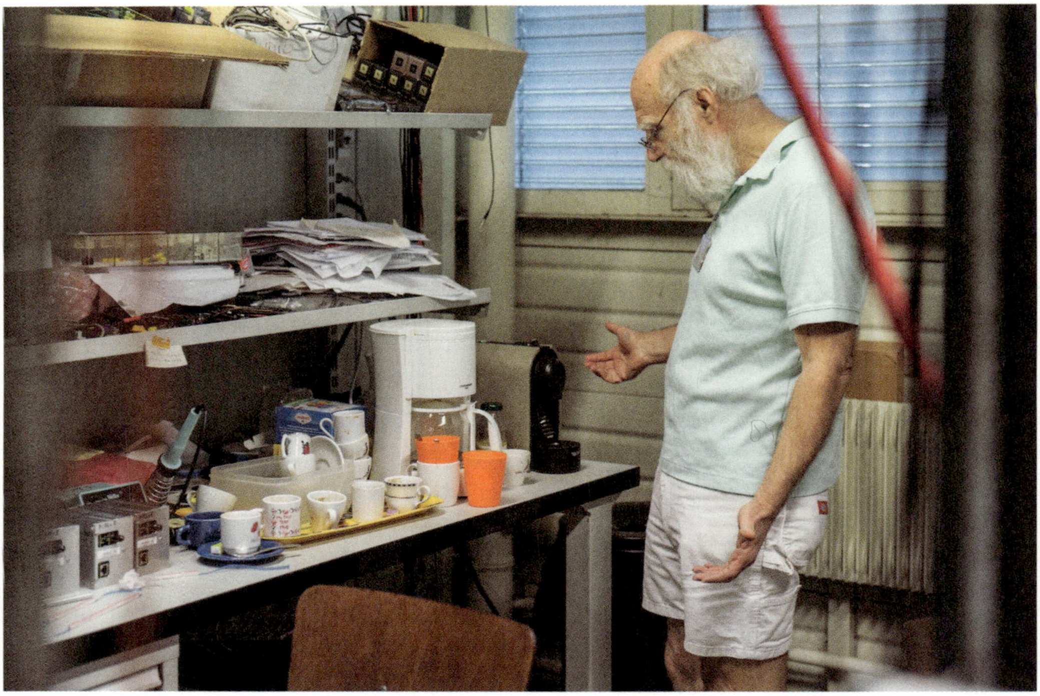

Fig. 3.9:
Andri Pol, *Menschen am CERN*, 2014. Photo © Andri Pol.

Rouben Mamoulian, 1931) or of the inventor fathoming the mysteries of electricity (*Metropolis*, dir. Fritz Lang, 1927) is increasingly ceding place in contemporary cinema to the image of the heroic sci-fi scientist pursuing research in a hermetic glass laboratory. The iconic attributes of the contemporary laboratory scientist are no longer a microscope and test tubes but glowing blue holograms (as, for example, in the lab portrayal in *Iron Man*, dir. Jon Favreau,

2008) or equally luminous blue screens, which are presented as the nerve centre of gigantic experimental set-ups. Such an image was created by director Ron Howard in his thriller *Angels and Demons* (2009), an adaptation of the eponymous bestselling novel by Dan Brown, which features the Geneva research complex CERN (European Centre for Nuclear Research) as a dramatic setting. Here, the nuclear scientists' laboratory is conceived as a transparent, open-plan zone with numerous screens and a huge glass front that affords a direct view of the Large Hadron Collider (LHC), an apparatus that has long since attained iconic status. The set thus suggests spatial proximity to the particle accelerator, although this would be inconceivable in real life, on account of radioactive and magnetic radiation. The reality in Geneva indeed looks very different: the campus is not dominated by a futuristic glass cathedral but by white concrete buildings constructed at various stages in the CERN's more than fifty-year-long history, while most scientists there work not at transparent control terminals but in cramped cubicles in buildings from the 1960s and 70s, frequently behind huge piles of photocopies and notes. This reality is the subject of a long-term documentation produced by the Swiss photographer Andri Pol, who accompanied people at the CERN as they went about their daily work (Pol 2014). Instead of superheroes in white coats, Pol reveals the nuclear scientists at CERN to be normal people in Bermuda shorts and hoodies, standing at the coffee dispenser, playing table tennis or lunching together beneath some trees (**Fig. 3.9**). As such, Pol's photographs counter the iconic view of the Large Hadron Collider with everyday human images and turn the ubiquitous myth of the CERN on its head. These scenes could not be further removed from the gigantesque fantasies of the sci-fi thriller. At least in the case of CERN nuclear scientists, there seems to be little chance that images from Hollywood will influence reality at any time soon.

REFERENCES

Benedict, Francis Gano. 1910. *Reports of Visits to Foreign Laboratories*. Vol. 2. Cambridge, MA: Harvard University Archives.

Benedict, Francis Gano. 1913–16. *Reports of Visits to Foreign Laboratories*. Vol. 3. Cambridge, MA: Harvard University Archives.

Dierig, Sven. 2002. "Apollo's Laboratory." http://vlp.mpiwg-berlin.mpg.de/pdfgen/essays/art8.pdf (accessed 10 February 2015).

Drösser, Christoph. 2013. "'Breaking Bad' Stimmt die Chemie?" *Die Zeit*, September 28. http://www.zeit.de/2013/40/breaking-bad-meth-chemie-wissenschaft (accessed 14 January 2015).

Galison, Peter and Caroline A. Jones. 1999. "Factory, Laboratory, Studio: Dispersing Sites of Production." In *The Architecture of Science*, edited by Peter Galison and Emily Thompson, 497-540. Cambridge, MA: MIT Press.

Gibson, Charles R. 1906. *The Romance of Modern Electricity*. London: Seeley.

Gooday, Graeme. 2008. "Placing or Replacing the Laboratory in the History of Science?" *Isis* 99 (4): 783-95.

Hill, Christopher R. 1975. "The Iconography of the Laboratory." *Ambix* 22 (2): 102-10.

Kuhn, Nicola. 2014. "Fotograf Thomas Struth im Interview. Es wird kalt in unseren Städten." *Der Tagesspiegel*, March 31. <http://www.tagesspiegel.de/kultur/fotograf-thomas-struth-im-interview-es-wird-kalt-in-unseren-staedten/9685046.html> (accessed 16 June 2015).

Neswald, Elizabeth. 2010. "An American Physiologist Abroad: Francis Gano Benedict's European Tours." <http://vlp.mpiwg-berlin.mpg.de/pdfgen/essays/art77.pdf> (accessed 10 January 2015).

Pickering, Andrew. 1995. *The Mangle of Practice: Time, Agency, and Science*. Chicago: University of Chicago Press.

Pol, Andri. 2014. *Menschen am CERN: Europäische Organisation für Kernforschung*, edited by Lars Müller. Zurich: Lars Müller Publishers.

Schmidgen, Henning. 2008a. "Wissenschaft: Das Labor als Archiv und Maschine." In *Poststrukturalistische Sozialwissenschaften*, edited by Stephan Möbius and Andreas Reckwitz, 450-66. 1st ed. Frankfurt / Main: Suhrkamp.

Schmidgen, Henning. 2008b. "Münsterberg's Photoplays: Instruments and Models in His Laboratories at Freiburg and Harvard (1891-1893)." http://vlp.mpiwg-berlin.mpg.de/pdfgen/essays/art71.pdf (accessed 10 February 2015).

Schmidgen, Henning. 2011. "Labor." *Europäische Geschichte Online*. http://www.ieg-ego.eu/de/threads/crossroads/wissensraeume/henning-schmidgen-labor (accessed 10 February 2015).

Seifer, Marc. 1998. *Wizard: The Life And Times Of Nikola Tesla*. New York: Citadel.

Tesla, Nikola. 2007. *Colorado Springs Notes: 1899-1900*, edited by Aleksandar Marincic. New York: BN Publishing.

Wurtz, Adolphe. 1870. *Les Hautes Études Pratiques Dans Les Universités Allemandes*. Paris: Imprimerie Impériale.

“Most clean rooms are dungeons with no sunlight or outside view. We planned on having windows because we wanted to humanize working conditions inside the clean room.”

Antonio Castro Neto, Graphene Research Centre,
National University of Singapore

4

THE MYTH OF TRANSPARENCY

Christina Landbrecht

Visitors to the Media Lab at the Massachusetts Institute of Technology (MIT) use a ramp to access the Upper Lobby on the second floor, and from there their gaze is drawn to the two transparent cuboid spaces to its left and right. Each of these accommodates an open-plan, double-height workspace surrounded by office spaces on a peripheral mezzanine. Here, professors and other faculty members occupy glazed rooms which allow them to see work going on in the lab below—and to be seen (**Fig. 4.1**).

Glazed cubes featured in the catalogue of requirements given to the Japanese architect Fumihiko Maki in 2010, when he was commissioned to extend the existing Wiesner Building designed by I. M. Pei & Partners. They are evidently a reference to "The Cube" designed by Pei as well as a response to the wish of Media Lab founder and former director Nicholas Negroponte to improve communication both among students and faculty members.[1] The multi-functionality of the glazed cubes builds on a long-standing tradition at MIT. The first "Cube" at the Wiesner Building established itself as a zone of experimentation and social encounter and served various needs, from study to exhibitions, from dining to debate. Its open-plan design enabled various objectives and activities to be simultaneously pursued. Yet in terms of ambience, Maki's

1 The author acquired this information on a visit to Maki's architectural studio in Tokyo in July 2013.

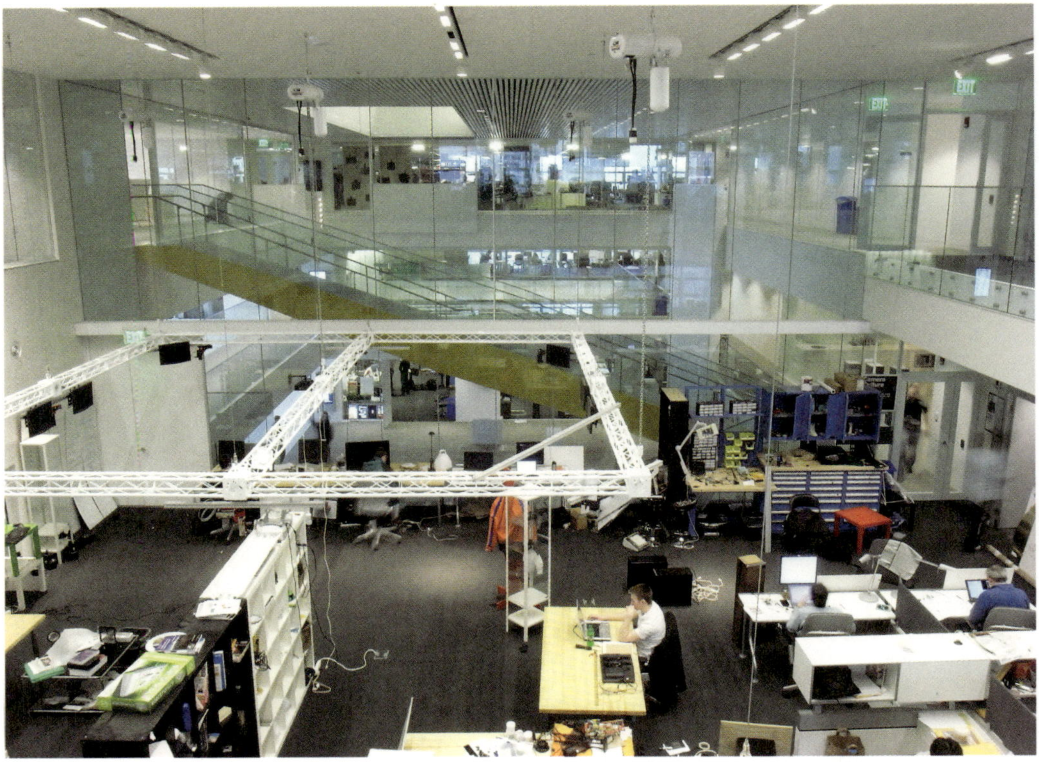

Fig. 4.1:
Maki and Associates, *Workspace*, MIT Media Lab, Cambridge, MA, 2007-9.
Photo © Tatsutomo Hasegawa.

cubes differ fundamentally from the original, which is a windowless hall, with a mezzanine and small offices along the lateral walls in which students work out of sight of any visitors (**Fig. 4.2**).

However, Maki's glazed version of "The Cube" was not unanimously welcomed. A student project with the title *Media Lab Transparency Garden,* which was displayed on the glass walls of the workshops in summer 2013, claimed the new design was created in the spirit of a "transparent turn". An info panel of the sort commonly found in museums described this move towards greater transparency as an economic tool of the kind already successfully deployed in the commercial sector to foster creativity, cooperation, decentralisation and the like. These values, it continued, were driving the current trend to transparency and openness and at the

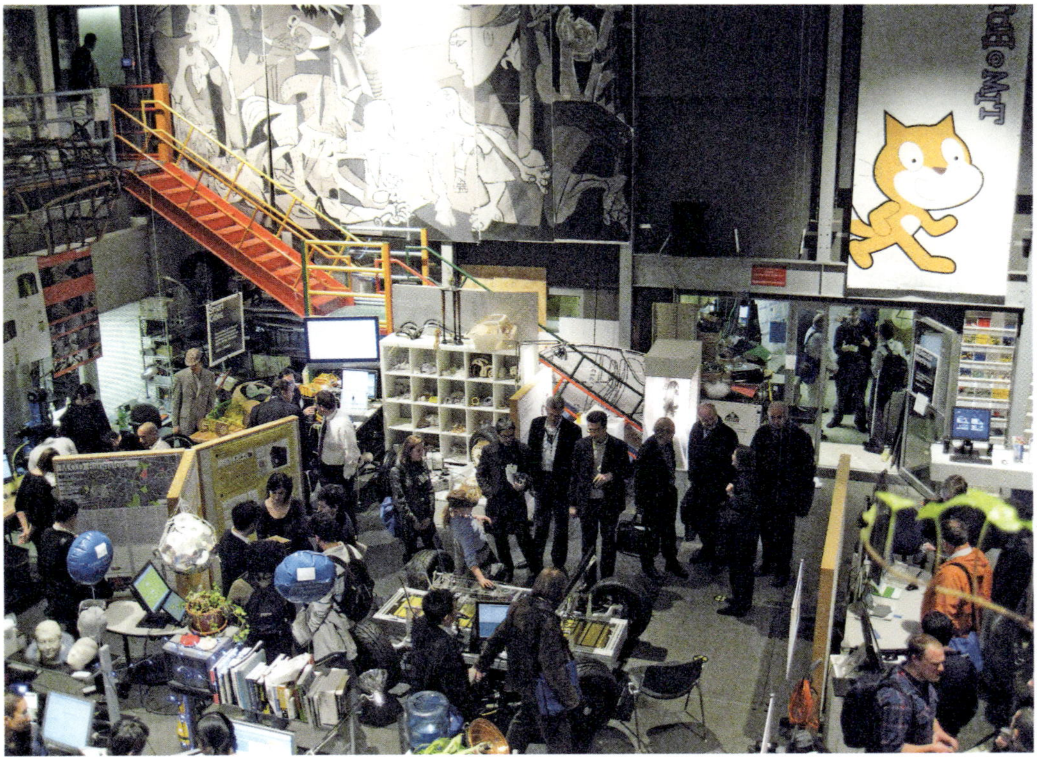

Fig. 4.2:
I. M. Pei & Partners, *The Cube*, The Wiesner Building / Center for Arts & Media Technology, MIT, Cambridge, MA, 1978-84. Photo © Zenovia Toloudi.

same time functioned as a means to save money and to profitably stage working procedures for public viewing: "These large, fish-bowl-like environments are cheaper than building offices and have the added benefit of providing a living spectacle for visitors to the Media Lab who can observe graduate students innovating in their native environment".

Since the turn of the twentieth century, the use of large-format glass has been not only a standard feature of modern architecture but also a source of controversy in recent architectural discourse (Foster 2011, Rowe and Slutzky 1968, von Moos 2002). Most controversial of all is the claim to transparency that generally accompanies the use of glazed façades (Vidler 1992, Sennett 1992). Yet despite such ambivalence, the glass laboratory is currently enjoying a boom. As the MIT Media Lab clearly demonstrates, the potent symbolism of glass is unabated, particularly with regard to new science buildings. One reason for this, as the following examples will demonstrate, is that laboratories today have a "communicative function" (Fenker 2014). They are not the hidden spaces they used to be but demonstrate both openness and transparency, which allegedly in turn foster productivity and innovation (Vasella 2009). Yet in many cases, this "insight" into modern laboratories is revealed to be merely a superficial gesture and one whose effect is often paradoxical: the view of the inside of the laboratory is generally conveyed solely by photographs of laboratory interiors while the laboratory itself remains an "opaque and concealed place" (Felsch 2005, 32).

HISTORICAL PRECEDENTS

Elements indicative of openness began to feature in laboratory architecture as early as the 1950s and 1960s: not only did glass play a substantial role here, but also a central plaza or atrium that heightened the impression of public accessibility. One prime example of the latter is the Salk Institute in California, which its Founding Director, the virologist Jonas Salk, commissioned from the architect Louis I. Kahn in 1959 (**Fig. 2.1** [See Chapter 2]). Salk's vision was of a building that would transcend merely practical considerations, an institute that would not only provide an optimal research environment but also meet researchers' personal needs. As he explained in an interview, "I thought that it would be well to consider establishing an institution that would be concerned not merely with nature, but with the human side of nature, not only with the molecular, cellular dimension, but what I call the human dimension" (Academy of Achievement 1991). When Khan presented his first blueprint for the Salk Institute in 1962, Salk therefore expressed "grave misgivings about the overall plan of the site, and the diffusion of the complex's social space into two separate but equal courtyards" (Leslie 2005, 100). In his opinion, only a central courtyard set between the two laboratory tracts could assure the social character of the building. The architect invested a great deal of time in elaborating this idea and finally acknowledged its importance: "One garden is greater than two because it becomes a place in relation to the laboratories and their studies. Two gardens were just a convenience. But one is really a place; you put meaning in it; you feel loyalty to it" (quoted in Treib 2006, 409). Hence the travertine-paved courtyard cut through by a small water channel came to forge a link between the two laboratory tracts flanking it. Yet the central courtyard was of importance to Salk not only because researchers had thus a shared space in which to exchange ideas and hence also a counterpoint to their usually solitary work at an individual desk or lab bench; it also allowed him to present the Salk Institute strategically as a space open to the general public—which was vital not least with regard to fundraising. Simultaneously, Salk strived to shield the work of the laboratories from public view: the structure of the laboratory tracts in combination with the "towers" (stacks of individual offices) fronting them proved an elegant solution.

However, the humanist and ambient concerns that shaped the design of the Salk Institute were not the only reasons laboratory buildings became transparent in the 1950s and 1960s. In Germany, the use of glass in architecture was read as a political signal, particularly after the Second World War. In 1958, Werner Heisenberg, then Director of the Max Planck Institute for Physics and Astrophysics in Munich-Freimann, commissioned a new building from the architect Sep Ruf (**Fig. 4.3**). This was the last building in the history of the Max Planck Institute to be built completely to the taste of the incumbent director. Heisenberg's personal decision to commission Ruf must be taken as a clear sign of his faith in what at the time was held to be avant-garde architecture (Zohlen 2007). In the 1950s, Ruf's buildings stood for a "disembodied" type of architecture: they were reviewed as steel and glass constructions of such transparency that they appeared to be in the process of dissolving (idem). Ruf's ensemble for the Academy of Fine Arts in Nuremberg (1952–54) and the German Pavilion that he and Egon Eiermann jointly designed for the World Exhibition of 1958 in Brussels are two cases in point. Both of these elegant, transparent, steel-skeleton, pavilion-style ventures won international acclaim. The journal *Architectural Review* even praised the German

Pavilion of 1958 as “the most sophisticated work of architecture in the exhibition” (quoted in Sigel 2000, 200).

Ruf’s plans for the Max Planck Institute—or, at least, for the auditorium and the façade of the experimentation hall—carried echoes of both. The result was a laboratory complex that was in striking contrast to the Max Planck Society’s earlier research

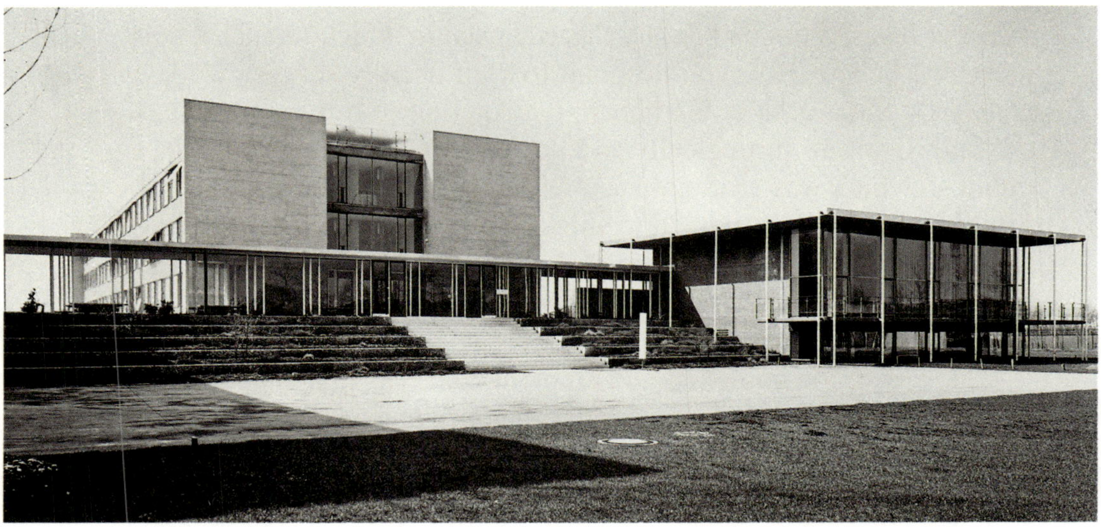

Fig. 4.3:
Sep Ruf, *Max Planck Institute for Physics and Astrophysics*, Munich, 1957-60.
Photo © Architecture Museum of the Technical University (TU) Munich.

buildings, which had been designed in the style of bourgeois villas. It was composed of loosely clustered horizontal units while the façade for the experimentation hall was nothing less than a milestone in the history of the Max Planck Society’s laboratory architecture: “In the main building [...] Ruf separated the façade from the actual structure and set it before it, like a membrane, while the glazed section he deployed for the first time ever here, at the Max Planck Institute, allowed one to surmise from outside the triadic structure, which is to say, the ground plan with a darkened lab section at its centre” (Braun 1999, 17).

That Ruf proposed to insert a central glazed section into the façade of the experimentation hall even though the lab spaces within it had to be protected from natural daylight underscores the symbolic value that both he and his client attributed to glass. The structure of the façade itself is distinctive, since the insertion of glass as a counterpoint to the pylon-like lateral elements dramatically enhances its transparency and apparent weightlessness. The striking portico—a motif Ruf had developed for the earlier projects in Nuremburg and Brussels—further heightened the impression that the experimentation hall was a publicly accessible complex. The elegant, clear portico thus served not only to attenuate the monumentality of the solid sections of the façade, but also to control the extent to which the laboratories were open to view—just like the arrangement of office stacks at the Salk Institute. These design decisions demonstrate the Janus-faced nature of the transparency aspired to at this particular Max Planck Institute: since the façade was only partially glazed, visitors were unable to see into the whole of the building. Additionally, the portico had an ambivalent effect as it drew the visitors to the transparent window in the façade yet also kept them at a distance. Ultimately then, the façade of the Max Planck Institute is more a symbolic gesture than a genuine invitation to step inside. It represents not so much structural accessibility as an attitude, which owes much to the reputation of the architect, who was lauded in 1954 as "one of the most influential theoreticians and lecturers in architecture of the 1950s in the Federal Republic of Germany" (Sigel 2000, 184). Ruf personified a democratic understanding of architecture, one that in the wake of the war and the Nazi regime was widely perceived as a symbolic stance against any buildings designed "to keep things secret" (Arndt 1961, 20).

FACILITATING COMMUNICATION IN THE LAB

In contrast to Heisenberg's symbolic appointment of Ruf or Salk's intervention in the design of the courtyard at La Jolla, the current trend to openness and transparency in contemporary lab architecture rarely depends on an individual's personal inclination or taste. Transparency now is part of the standard repertoire of science buildings. The trend is rooted, however, in more than purely aesthetic considerations and therefore has more far-reaching repercussions than in the 1950s and 1960s. As, inter alia, the architect Michael Fenker has emphasised, contemporary science architecture must first and foremost rise to the challenge of facilitating communication (Fenker 2014). Yet this is the case, one must add, only in those fields of scientific activity that may be technically, morally or economically exposed, without hesitation, to public view. Laboratories in which experiments are carried out on animals, such as the Institut du Cerveau et de la Moelle Épinière (Institute for Brain and Marrow Research) in central Paris, are either located underground or, as in the case of the Laserzentrum Nord (Laser Centre North) in Hamburg, designed as windowless cubes, since the technologies in use there cannot be exposed to daylight. However, in the case of the workbench or the clean room laboratory type, transparency is now standard practice. This is justifiable on two grounds: firstly, glass walls are used internally in laboratory buildings so as to enhance transparency for the researchers working there, in the hope that this will foster more interaction and thus increase productivity. Secondly, glass façades allow anyone to look at will into the laboratory and thus respond at least superficially to demands for the democratisation of knowledge.

It must be asked, nonetheless, whether the widespread use of glass in laboratories really is effective, or whether perhaps the sociologist and philosopher of science Bruno Latour was right when he claimed that transparency tends to have a thoroughly negative impact on the performance capacity of science:

> Accurate facts are hard to come by, and the harder they are, the more they entail some costly equipment, a longer set of mediations, more delicate proofs. Transparency and immediacy are bad for science as well as for politics; they would make both suffocate.
> (Latour 2005, 21)

A look at the Molecular Foundry, a nanoscience research institute completed in 2010 in Berkeley, California, suffices to show that even a few well-placed design elements can lend a building an air of openness and transparency. Moreover, they enable the commissioning client and the architect to stage the laboratory in a way that assures it a strong public profile. A short clip created for the occasion of the Foundry's inauguration emphasises that the institute is opening up so as to make the research pursued there broadly available. The architect Bill Diefenbach, who led the project on behalf of SmithGroupJJR, explained that the terrace in particular served an important function: "on the south side [there is] a very sun-filled plaza, which is a green roof, very necessary to provide an interactive space that's outside—to bring the researchers outside" (SmithGroupJJR 2010, 2:15 min). He justifies the expensive roof terrace, claiming that it not only fosters internal communication between scientists working at the institute but also highlights the building's broader social purpose, inasmuch as it serves as a quasi-interface between science and the general public: "We also wanted to express the focus of the laboratory out into the community, because [the scientists] are here to create solutions for all of us" (idem, 2:23 min.). Yet the Molecular Foundry is located neither on a busy campus nor in an urban residential neighbourhood, but thrones rather in solitary hilltop splendour, over San Francisco's Bay Area. Anyone wanting to benefit from its alleged public accessibility first has to work out how to reach it. The architects, Diefenbach and his colleague Suzanne Napier, emphasise several times in the clip how they strived to design a lab that would contribute in a positive way to society. They turn the spotlight on the social relevance of the nanoresearch pursued at the Molecular Foundry as well as on the role architecture can play as a facilitator of communication.[2] In their view, a laboratory should not be conceived merely as accommodation for practical experimentation, but should also serve purposes of communication. At the Foundry, however, only the roof terrace fulfils this function. Laboratory spaces within the building—many of which, as the clip shows, require special lighting—generally cannot be built of glass.

2 It is to be noted here that the Molecular Foundry was designed as an open-access lab and nanoscience user facility and thus has a more open structure than most other labs. Scientists from all over the world are able to work in this lab and to use both the apparatus and the expertise of the scientists employed there in order to further their work in the field of nanosciences.

Thus the roof terrace alone stands—not only in the film—as an emblem of the laboratory's openness and accessibility.

A similar function falls to the glazed in-between level of the :envihab [environmental habitat] at the Cologne–Porz branch of the DLR, the national aeronautics and space research centre of

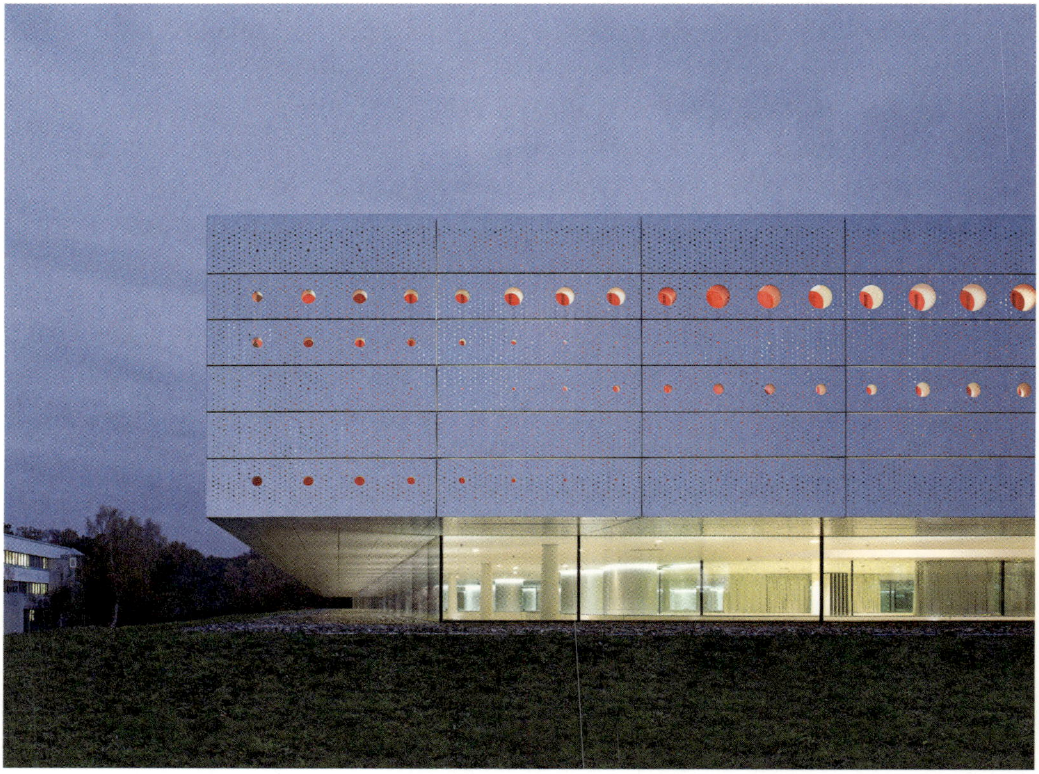

Fig. 4.4:
Glass Kramer Löbbert and Uta Graff, *:envihab—Future Research for Space and Earth*, Cologne, 2009-13. Photo © Christian Gahl.

the Federal Republic of Germany. The :envihab is a laboratory in which both living and research conditions in outer space are examined by means of simulation. It had to be designed as a self-contained space independent of its actual climatic environment and to be built as a dual-level, partly subterranean building. Yet despite this profile there were calls for it to be a publicly accessible space

(TU Munich 2014). Glass Kramer Löbbert and Uta Graff, the architects of the building, proposed that a completely glazed level be inserted in-between the deep roof structure (in which all the requisite technology is installed) and the underground laboratories; for this, as Johannes Löbbert said, "would arouse the curiosity of passersby" (Löbbert 2014).

A look into the in-between level reveals one thing above all: numerous so-called media tubes of various diameter randomly distributed in the open-plan hall serve to link the technology installed in the roof with the equipment installed on the underground level (**Fig. 4.4**). This, of course, is disappointing for potential visitors. Instead of being able to grasp the laboratory as a spatial entity or even to observe the experimental work being conducted within it, there is nothing to be seen except a number of enclosed media tubes which could easily be mistaken for pillars serving a purely structural purpose. This essentially reduces the laboratory to its technical and structural dimension. It is abstracted to such an extent that it has become a "black box"—a term which Bruno Latour chose to describe something which "acts as one piece [...] no matter how many pieces [...] are in it and no matter how complex the [...] system" (Latour 1987, 131).

LABORATORIES WITHOUT WALLS

More than ever before, universities and companies commissioning laboratory buildings now place great value on public accessibility (Gieryn 2008, 797) and on architectural designs that foster internal communication between research scientists (Vasella 2009). According to the sociologist Thomas F. Gieryn, one good example if not paragon of this trend is the James H. Clark Center for Biomedical Engineering and Sciences at Stanford University (See Chapter 5), which was designed by Norman Foster & Partners in cooperation with MBT Architecture (now part of the architecture office Perkins + Will):

> The design of the Clark Center is itself an experiment: can breakthroughs and innovations be nurtured most effectively in

> a building where the spatial locations of people and equipment can be rearranged at any time, in response to rapidly changing patterns of transdisciplinary interaction and collaboration? (Gieryn 2008, 797–98)

The three convex and completely glazed laboratory wings of the Clark Center projecting into a central courtyard offer staff and visitors an invitingly fluid laboratory landscape (**Fig. 2.2** [See Chapter 2]). All the furnishings, even workstations, are on wheels. Continuous exterior balconies obviate the need for interior corridors and also offer curious passersby a view of the experiments going on inside the building. The Center's laboratories can thus be easily and rapidly modified: benches and desks can be moved in a way "that allows ad-hoc team formations in response to fast-evolving research needs" (Foster 2003). The Clark Center is striking not only on account of its experimental interior design but also because of how it is projected into public space. The open-plan laboratory space is completely visible to anyone entering the courtyard and clearly communicates an innovative spatial concept:

> The iconic form of Clark Center embodies the collaborative spirit of Bio-X. Sweeping windows reveal open lab spaces which dissolve the walls between labs, people and ideas. Researchers flow through this nexus that blends traditional departments, schools and areas of study. (Stanford Bio-X 2013)

The fluent transition from laboratory to public space would be less spectacular had the lab not been set at an almost central location, as the showpiece and pulse of the campus. Its strategic position and the blurring of borders between interior and exterior space serves to highlight the Clark Center's role, not only as an innovatively designed interdisciplinary research building that brings together engineers, chemists, physicists, medical doctors and biologists, but also—as one can read in a brochure published by one of the architect offices behind the project—as a "social magnet for the University" (Foster 2003). Instead of marking the laboratory as "a building set apart" (See Chapter 2), the Clark Center figures as a quasi public space. Indeed the laboratory appears so accessible and welcoming to visitors that, as Gieryn reports, scientists working there have had to try hard to re-establish the exclusive ambiance of their research environment by pasting on its glazed façade handwritten notices announcing: "Experiments in Progress—No Public Tours" or, more explicitly, "Please do NOT ask to open the door!!!" (Gieryn 2008, 798)

In the light of this, it is no surprise to note that the radical transparency epitomised by the Clark Center is not universally regarded as a viable model, despite Gieryn's claims to the contrary. Even though this same aspiration to public accessibility came under consideration, for instance, during the comprehensive construction programme carried out in Basel on behalf of the Swiss pharmaceutical company Novartis AG, it was ultimately drastically revised. For while the glass laboratories at Novartis were meant to assure a pleasant working atmosphere and foster creative exchange among scientists (Buchmüller et al. 2009, 203), the public is not permitted to observe work in progress there. The Italian architect and Professor of the History of Urban Planning at the ETH Zurich, Vittorio Magnago Lampugnani, was commissioned to draw up the master plan for the company site, which eventually came to include laboratories and offices designed by leading contemporary architects such as Herzog & De Meuron, David Chipperfield, Tadao Ando and Frank O. Gehry (See Chapter 5). In the meantime, the campus is open on certain Saturdays to interested visitors, who are able to see the site itself, its specially commissioned artworks and façade designs, as well as certain lobbies and atriums in the research buildings. For Novartis—as the obligatory guided tours without which one cannot access the site make very clear—transparency is the new company philosophy. As at the Clark Center, it is believed to help improve communication structures and hence optimise conditions for interdisciplinary research. In line with the company's philosophy, the design of a laboratory is taken to have a decisive impact on research productivity. According to the biochemist Gottfried Schatz, the research scientists' spokesman: "The greatest architectural threat to a research institution lies not in a building's façade, but in the walls that subdivide its interior" (Schatz 2009, 14–15).

At the Clark Center, maximum visibility led to the research scientists' wish to be able to pursue their work undisturbed. At Novartis, this is unlikely ever to be the case. Although transparency is writ large, visitors to the campus are actually denied access: outsiders are never able to look inside the laboratories (See Chapter 5). In contrast to the Clark Center, where floor-to-ceiling windows serve to stage scientific work for public viewing, transparency at Novartis is revealed to consist in nothing more than a range of visual formulas. Only the reception pavilion designed by the architect Marco Serra and built from 2004 to 2007 at the gateway to the campus is completely transparent. Despite its size, it has neither a steel nor reinforced concrete skeleton but rather a façade composed of U-shaped elements comprised of glass fins and large panes of glass. The minimalist and seemingly free-floating office building realised from 2004 to 2006 by the Japanese architectural practice SANAA and situated near the reception pavilion likewise has a completely glazed façade that permits a view of its interior. It is accordingly described to participants in the Novartis guided tours as "a doll's house"—which proves that the architects' use of delicate supports and extensive glazing met the demand for transparency. However, it is one of the few buildings on the campus in which there are no laboratories, since it serves purely administrative purposes. A look inside the campus buildings, if permitted at all, is confined to office spaces, like those designed by SANAA and also Gehry, which are entirely transparent but, significantly, not a laboratory.

Another case in point is the laboratory building designed by David Chipperfield. It demonstrates that claims to transparency on the Novartis campus should not be taken to imply public access to work in progress in the laboratories. Here, the glass laboratory is nothing more than an image hung in the lobby, namely a work with the title *Lab* by the Berlin-based artist Menno Aden, which won him the German Prize for Scientific Photography in 2013 (**Fig. 4.5**). It presents a view of one of the building's most spacious and innovatively designed laboratories within which mobile workbenches with integrated gas supply and other utilities have been arranged like letters of the alphabet or numbers on a clock face. Yet anyone who thinks, on seeing this image, that he or she has finally

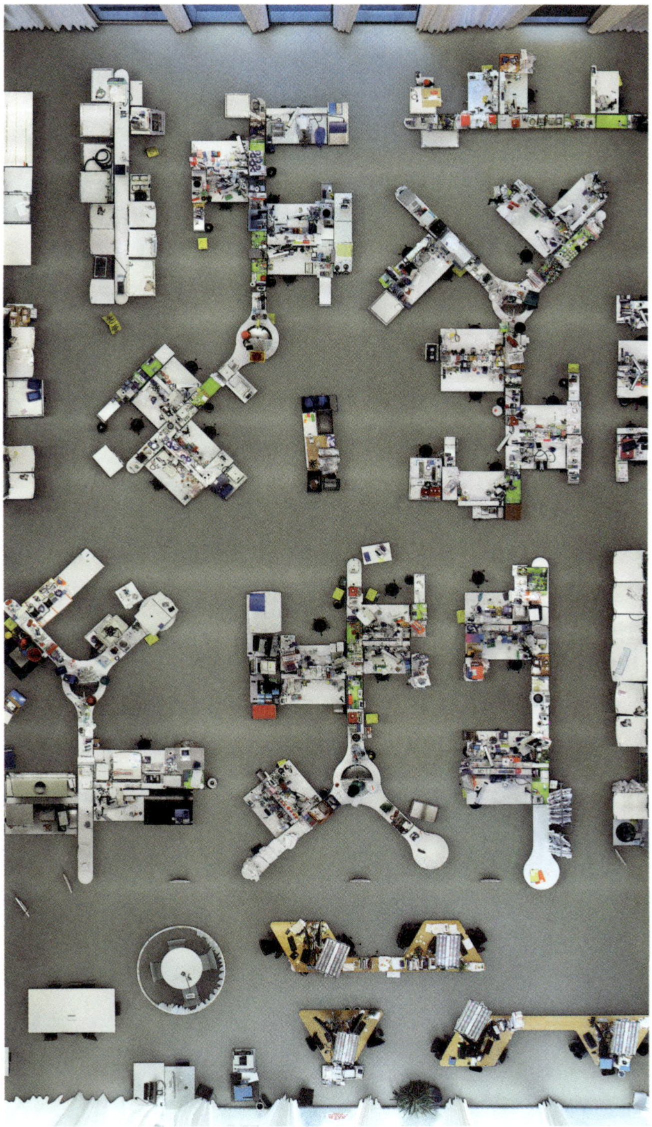

Fig. 4.5:
Menno Aden, *Lab*, photograph, 2013, 261×149 cm. Photo © Menno Aden.

found an authentic view of a campus laboratory will be quickly disappointed. For this is not a documentary photograph but rather a digital composition of several images. The artwork thus does not afford a view of an actual laboratory inside the Chipperfield building but only of a laboratory fiction. It reveals the Novartis transparency credo to be paradoxical in much the same way as the crystalline structure of the neighbouring laboratory building designed by Yoshio Taniguchi does. For, thanks to foils, the latter in the daytime is an opaque white cube; only at night, when public traffic has ebbed away, does it reveal its interior life.

As these examples show, the role of communication facilitator, which Fenker maintains modern laboratories must take into account, has certainly been an important component of the design briefs given to architects since the turn of the millennium—but the transparent laboratory nonetheless remains elusive.

During the lecture series *The Theater of Proof*, which accompanied the exhibition *Laboratorium* in 1999, Bruno Latour stated that if ever the public's awareness of scientific procedures were to be heightened then laboratories must be opened up. In contrast to the scepticism he expressed in 2005 regarding transparency in science, Latour argued then that society had entered into an "experimental age" in which the experiment has an impact on human beings and that not only political awareness but also scientific awareness is therefore vital: "In the past only scientists needed an understanding of the internal functioning of the laboratory, while the lay public had to be satisfied with the finished results. Today, this is no longer the case, for there is now a growing interest in understanding how experiments work" (Latour 2001, 185). Despite its promises and gestures, architecture has not yet given the general public reason to hope it might soon gain insight into the work of laboratories. If, however, we believe the images that are circulated, this need appears to have already been met.

THE PICTURE OF TRANSPARENCY

Although actual insight into laboratories is mostly denied, transparency remains an important pictorial motif, one conveyed above all by documentary images of laboratory interiors taken upon completion of university buildings and science and technology centres. Generally commissioned for advertising or publication purposes by major architectural practices, such as Payette Associates, Perkins +

Will, SmithGroupJJR or HDR, they have served to establish an image of the laboratory that is of crucial importance to the public's current understanding of scientific knowledge.

As one quickly realises upon reviewing the available imagery, the recurrent iconographic idiom has been condensed from very few elements. As a rule, the life sciences workbench laboratory most commonly featured in images of lab interiors is presented against the backdrop of a natural, park-like landscape. Either clusters of trees in autumn hues are the foil in the background

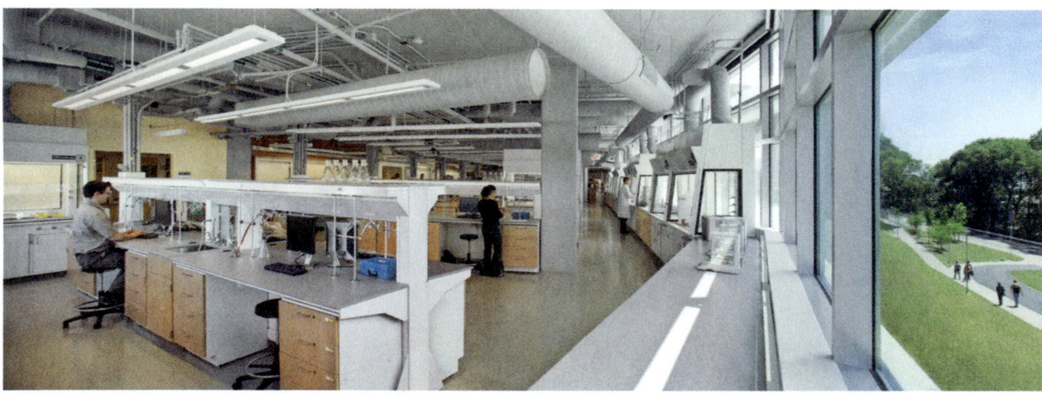

Fig. 4.6:
Payette Associates, *Labspace Interior*, Carl J. Shapiro Science Center, Brandeis University, Waltham, MA, 2009. Photo © Warren Jagger Photography.

or swathes of parkland are reflected in the glazed front façade. Laboratories thus appear to be spacious, airy and open to their surroundings (**Fig. 4.6**). Verdant nature is incorporated above all in contemporary life science laboratory landscapes, perhaps because nature seems the perfect backdrop to a discipline concerned with research into natural objects. In striking contrast to the greenery, the gleaming white laboratory emits an aura of meticulous order and cleanliness, only occasionally enhanced by timber detailing on drawers or black work surfaces. This focus on the white look of the laboratory is often further underscored by the presence of lab users, who generally pose in white coats. In contrast to those who populate conference rooms, lobbies, stairwells and databank rooms, the scientists pursuing research in laboratories wear white coats—and often gloves and protective glasses, too—virtually as a symbol of their profession. Their clothing not only signals their status as experts but also allows them to blend into the background, almost as if they were a part of the furnishings (**Fig. 4.7**). Moreover, photographs of researchers at work without a white coat tend to highlight the relationship between them and their equipment. Here colour analogies play a pivotal role: for example, the yellow T-shirt worn by a research scientist may match the lower part of the scales he or she is using (**Fig. 4.8**), or the shade of someone's shirt may resemble the workbench on view (**Fig. 4.6**).

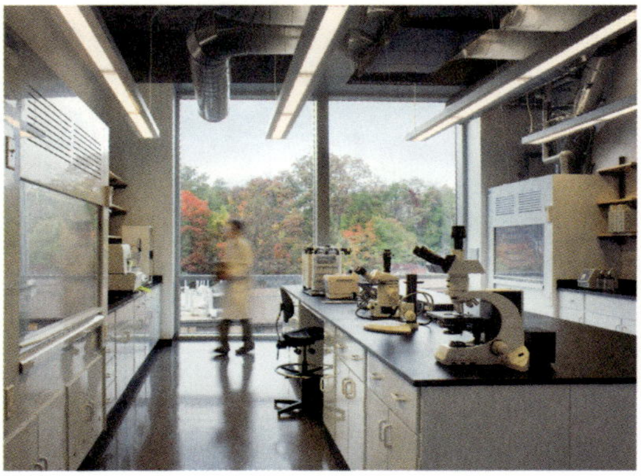

Fig. 4.7:
Payette Associates, *Labspace Interior*, Gary C. Comer Geochemistry Building, Columbia University, New York, 2006-7. Photo © Warren Jagger Photography.

This harmonised depiction of the scientist and his or her equipment is more than simply an iconographic idiom. To link persons and things in this way fundamentally characterises the work of the laboratory, as Knorr Cetina has emphasised:

> Not only objects but also scientists are malleable with respect to a spectrum of behavioral possibilities. Moreover, it is not at all clear that these scientists must remain stable individual entities that are separated out from other objects in the

> laboratory. Certain of their features may be coextensive with those of objects; they may be construed as 'coupled' to objects and machines or they may 'disappear,' as individual players in epistemic collectives that match the objects in the laboratory. (Knorr Cetina 1995, 146)

The image perpetuated of the laboratory today has little in common with that of the creative chaos to be found in genre portraits of either the alchemist or the inventive genius amid bubbling systematic experiments (See Chapter 3). Rather, the contemporary laboratory is presented as a technical, artificial and clean think

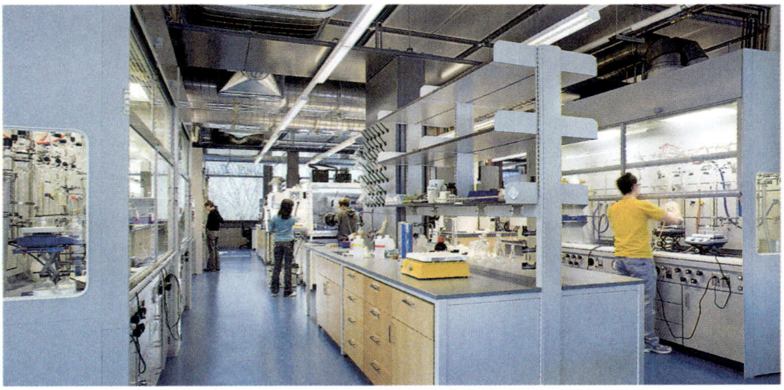

Fig. 4.8:
Hopkins Architects & Payette Associates, *Labspace Interior*, Frick Chemistry Laboratory, Princeton University, New Jersey, 2007-10.
Photo © Warren Jagger Photography.

tank. It is a space that has undergone a radical transformation since the days of cluttered, cramped and provisional set-ups, such as the shed in which the Curies isolated radium. The imperative nowadays is not isolation but openness, not makeshift means but a comfortable working environment. Today's laboratory is therefore even often defined as "beautiful", as an interview with senior staff at Novartis made clear. In answer to a remark made by Simon Heusser, the Senior Strategic Communications Advisor at Novartis AG—"[that a] glance back in time shows that numerous discoveries were made in working environments to which one would hardly dedicate a volume of essays"—Mark C. Fishman, President of the Novartis Institutes for BioMedical Research, and Jörg Reinhardt, Chairman of the Board of Directors at Novartis AG, both replied that "beauty motivates people and it helps to recruit people in the competitive work situation of today" (Fishman and Reinhardt 2009, 34).

An open and transparent space thus not only promotes mutual exchange and interaction among scientists, but also perpetuates an image of their work that they are happy to convey to the public. The image of the laboratory as a neat space with an atmosphere of concentration, where scientists are as likely to be found engaged in group discussions as sunk in solitary contemplation of their experiments, illustrates that which the report *Research & Research Facility Futures: 2012–2014 and Beyond* predicted for the future of research: "Research activity will become an increasingly social process involving face-to-face collaboration within small interdisciplinary teams focused on targeted research problems" (Tradeline 2012). Such images are decisive for our current notions of what is meant by "knowledge production" in the laboratory setting. The sociologist Nico Stehr has pointed out that studies undertaken by social scientists on the theme of knowledge have been relatively rare to date. He concluded that "knowledge is understood to be a 'black box'" (Stehr 1994, 203). The view into the glass laboratory creates the impression that the opposite is true, an impression due first and foremost to the fact that increased public access to places of experimental practice has transformed the traditional notion of the laboratory as a locus of "hermetic knowledge production" (Felsch 2005, 32). This is unlikely to change in the near future, as long as knowledge remains a "black box", as long as the general public is denied a look into the laboratory and as long as transparency is more an image than a reality.

REFERENCES

Academy of Achievement, 1991. "Jonas Salk Interview—The Calling to Find a Cure." Last modified October 28, 2014. <http://www.achievement.org/autodoc/page/sal0int-1> (accessed 9 March 2015).

Arndt, Adolf. 1961. "Demokratie als Bauherr." In *Anmerkungen zur Zeit*. Vol. 6. Berlin: Akademie der Künste.

Braun, Hardo, Dieter Grömling, Carl-Egon Heintz, and Alfred Schmucker. 1999. *Bauen für die Wissenschaft: Institute der Max-Planck-Gesellschaft*. Basel: Birkhäuser.

Buchmüller, Albert, Robert A. Ettlin, Gaby Keuerleber-Burk, Martin Kieser, Roger Müller, Reto Naef, Markus Oser, and René Rebmann. 2009. "Implementation and Management Strategies." In *Novartis Campus: A Contemporary Work Environment: Premises, Elements, Perspectives,* edited by Jaqueline Burckhardt, Martine Francotte, Vittorio Magnago Lampugnani, Michael Renner, Wolfdietrich Schutz, 192-209. Ostfildern: Hatje Cantz.

Fishman, Mark C. and Jörg Reinhardt. 2009. "The Construction of Space for Research and Development." In *Novartis Campus: A Contemporary Work Environment: Premises, Elements, Perspectives,* edited by Jaqueline Burckhardt, Martine Francotte, Vittorio Magnago Lampugnani, Michael Renner, Wolfdietrich Schutz, 34-37. Ostfildern: Hatje Cantz.

Felsch, Philipp. 2005. "Das Laboratorium." In *Orte der Moderne: Erfahrungswelten des 19. und 20. Jahrhunderts*, edited by Alexa Geisthövel and Habbo Knoch, 27-36, 1st ed. Frankfurt / Main: Campus Verlag.

Fenker, Michael. 2014. "Laborbau international: Laborbau in Frankreich." Paper presented at the 10th Laborrunde conference, Berlin, May 8-9.

Foster, Hal. 2012. "Crystal Palaces." In Hal Foster. *The Art-Architecture Complex*. 34-51. London: Verso.

Foster & Partners. 2003. "Clark Center, Stanford University—Interiors." <http://www.fosterandpartners.com/design-services/interiors/clark-center-stanford-university-interiors/> (accessed 10 February 2015).

Gieryn, Thomas F. 2008. "Laboratory Design for Post-Fordist Science." *Isis* 99 (4): 796-802.

Knorr Cetina, Karin. 1995: "Laboratory Studies: The Cultural Approach to the Study of Science." In *Handbook of Science and Technology Studies*, edited by Sheila Jasanoff, Gerald E. Markle, James C. Peterson, and Trevor Pinch, 140-66. London: Sage Publications.

Latour, Bruno. 1987. *Science in Action: How to Follow Scientists and Engineers Through Society*. Cambridge, MA: Harvard University Press.

Latour, Bruno. 2001. "The Theater of Proof: A Series of Demonstrations." In *Laboratorium*, edited by Hans Ulrich Obrist and Barbara Vanderlinden, 185-87. Köln: Dumont.

Latour, Bruno. 2005. "From Realpolitik to Dingpolitik or How to Make Things Public." In *Making Things Public: Atmospheres of Democracy*, edited by Bruno Latour and Peter Weibel, 4-31. Cambridge, MA: MIT Press.

Leslie, Thomas. 2005. *Louis I. Kahn: Building Art, Building Science*. New York: George Braziller.

Löbbert, Johannes. 2014. ":envihab am DLR in Köln: Ein Bau zur Erforschung der Zukunft für Weltraum und Erde." Paper presented at the 10th Laborrunde conference, Berlin, May 8-9.

Rowe, Colin and Robert Slutzky. 1968. *Transparenz*. Basel: Birkhäuser.

Schatz, Gottfried. 2009. "A Birthplace for New Ideas." In *Novartis Campus: A Contemporary Work Environment: Premises, Elements, Perspectives,* edited by Jaqueline Burckhardt, Martine Francotte, Vittorio Magnago Lampugnani, Michael Renner, Wolfdietrich Schutz, 12-19. Ostfildern: Hatje Cantz.

Sennett, Richard. 1992. "The Technology of Unity." In Richard Sennett. *The Conscience of the Eye: The Design and Social Life of Cities*, 103-11. New York: W.W. Norton & Co, Inc.

Sigel, Paul. 2000. *Exponiert: Deutsche Pavillons auf Weltausstellungen*. Berlin: Verlag Bauwesen.

Smith Group JJR. 2010. "Lawrence Berkeley National Laboratory Molecular Foundry: A Center for Nanotechnology Research." *You Tube* video, 4:14. February 23. <https://www.youtube.com/watch?v=s3eymjyPiFM> (accessed 10 February 2015).

Stanford Bio-X. 2013. "The James H. Clark Center." https://biox.stanford.edu/about/clark-center (accessed 9 March 2015).

Stehr, Nico. 1994. *Arbeit, Eigentum und Wissen: Zur Theorie von Wissensgesellschaften*. 1st ed. Frankfurt / Main: Suhrkamp.

Tradeline. 2012. "The Growing Human Factor in Research." Last modified January 16, 2012. <https://www.tradelineinc.com/reports/2012-1/research-research-facility-futures> (accessed 19 March 2015).

Treib, Marc. 2006. "To End a Continent: The Courtyard of the Salk Institute." *Journal of the Society of Architectural Historians* 65 (3): 402-27.

TU Munich. 2014. "Fertigstellung Neubau :envihab—Institut für Luft- und Raumfahrtmedizin." <http://www.ar.tum.de/aktuell/news-singleview/article/fertigstellung-neubau-envihab-institut-fuer-luft-und-raumfahrtmedizin> (accessed 9 March 2015).

Vasella, Daniel. 2009. "Foreword." In *Novartis Campus. A Contemporary Work Environment: Premises, Elements, Perspectives*, edited by Jaqueline Burckhardt, Martine Francotte, Vittorio Magnago Lampugnani, Michael Renner, Wolfdietrich Schutz, 8-9. Ostfildern: Hatje Cantz.

Vidler, Anthony. 1992. "Transparency." In Anthony Vidler. *The Uncanny: Essays in the Modern Unhomely*. 217-25. Cambridge, MA: MIT Press.

Von Moos, Stanislaus. 2002. "'Glanzblitz!' Über Architektur, Transparenz und Multimedialität." In *Material in Kunst und Alltag*, edited by Dietmar Rübel and Monika Wagner, 47-63. Berlin: Akademie-Verlag.

Zohlen, Gerwin. 2007. "Der Henry Ford Bau im Kontext der Nachkriegsmoderne." Paper presented on the occasion of the re-opening of the Henry Ford Building at the Free University Berlin, April 16. <http://www.fu-berlin.de/sites/hfb/downloads/zohlen_vortrag.pdf?1307221934>. (accessed 19 March 2015).

FUNCTION AND DESIGN

“In my experience, there appears to be a universal design for the structure and organization of laboratories. At least in chemistry, to have laboratory benches, hood space and instrumentation space seems to be the norm.”

Steven L. Bernasek, Department of Chemistry, Princeton University

5

BETWEEN SERVICE AND REPRESENTATION: SCIENTIFIC ARCHITECTURE SINCE THE MID-TWENTIETH CENTURY

Sabine Hansmann

In the mid-1980s, when the American architect Robert Venturi set about planning a new molecular biology laboratory building for Princeton University, he was faced with the challenge that arises with every architectural project: to strike a balance between generic demands and specific in-situ conditions. Venturi felt that the former carried the most weight: "The most important thing, in my opinion, about the architecture of the Research Laboratory is that it is *generic*" (Venturi 1999, 388). And, he continued, this holds true in particular for the individual laboratories within a research institute, as these must be as flexible, functional and neutral as possible.

Flexibility is a fundamental paradigm in laboratory architecture to this day. The trend to so-called open labs became manifest in most research institutes by the mid-1990s. Continuous, open-plan zones with large, communal areas marked a striking departure from the traditional individual "closed" laboratory (Watch 2001, 9–11). Generally, the new labs are designed for use by several researchers or research teams simultaneously and, as a shared resource easily adapted to a range of needs, they significantly reduce overall costs. They also foster impromptu encounters between research scientists. The open lab typology is particularly advantageous for interdisciplinary teams engaged in projects of limited duration, for it can quickly accommodate shifts in thematic focus and staffing requirements. Flexible arrangements are increasingly important, moreover, not only in the laboratory itself but also in the overall structural design of a research institute. An open-plan building with a modular wall system allows not only the size of a space but also its purpose to be altered to meet changing needs. Laboratory workstations are now mostly designed as standardised modules while utilities—the general infrastructure—are distributed evenly throughout an entire lab area. This allows each workstation to be positioned as required and equipped for the tasks in hand via plug-in systems (**Fig. 5.1**). Thus not only the

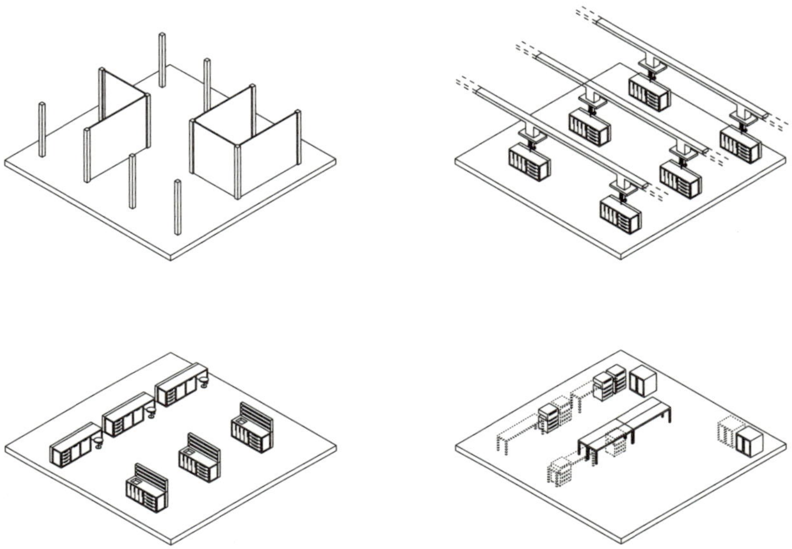

Fig. 5.1:
The laboratory: a systematic space. Drawing © Patryk Kujawa.

furnishings in a laboratory but also the very space itself have been systematically refined so as to assure the highest possible degree of flexibility (Hegger 2005; Watch 2001, 3–20; Clynes and Branton 1984).

The open-plan zone developed according to concerns of usability as well as adaptability to a range of different needs. It constitutes a particularly successful form of the generic architecture that Venturi called for. However, in the eyes of the historian of science Robert E. Kohler, that which makes the laboratory generic is not so much its spatial flexibility as the evident uniformity of its design. He notes, moreover, that "[g]eneric places sustain the illusion that their inhabitants' beliefs and practices are everyone's beliefs and practices. We credit knowledge and practices that are universal and mistrust what is merely local and particular, and laboratories are meant to be universal, the same everywhere" (Kohler 2002, 7). Standardisation was indeed a major feature of the so-called laboratory revolution that took place in the latter third of the nineteenth century (See Chapter 2). It culminated in the early 1950s in the modular, flexible laboratory still with us today (Grube 2005), the uniformity of which makes it an instantly identifiable motif.

However, international competition for top research scientists and a cutting-edge reputation means that research institutions increasingly opt for something more than an adaptive and utilitarian laboratory building. One might echo Kohler in arguing that the generic nature of laboratory architecture guarantees the research conducted within it greater credibility. Yet in response to ever stiffer global competition, the need for a location with specific qualities has become increasingly pressing (Hegger 2005, 31), particularly given research scientists' limited institutional affiliation and high degree of international mobility (Stichweh 1996). Research institutions are accordingly often built not only with functionality in mind but also at a location likely to enhance the distinctive profile of the commissioning university or company. Entire buildings or particularly prominent sections are increasingly created as objects of prestige. This is especially important in the international context, as William James, Pro Vice-Chancellor for Planning and Resources at Oxford University, has emphasised:

> Everybody in Oxford understands why it is one of the best places in the world to work, but not everyone outside the UK understands that, so potential top researchers and scientists need convincing at a visual/visceral level. We need to get people to think 'I can see myself working there'. […] It really makes a difference in recruitment and retention of staff. The perfect building is not enough to deliver good science, but it sure makes it easier.
> (Laboratory News 2015)

Architecture now serves as a showcase in order to appeal to renowned scientists from all over the globe as well as to potential partners, funding bodies and sponsors (Wentzlaff et al. 2014; Laboratory News 2015).

Thus there is a need, on the one hand, for a generic laboratory space equipped with standardised modular systems that provide research scientists with the most functional and neutral working environment possible; and, on the other, for iconic buildings of the sort required by research institutions intent on cementing a strong and unmistakable public profile. The few historical and contemporary examples explored below serve to demonstrate how architecture may resolve the tension between these conflicting demands: for a generic space and for iconic buildings that ensure a particular location.

OPEN SPATIAL SYSTEMS

The evolution of the modern research building, with its typically high degree of structural flexibility, mutability and adaptability, was propelled in the United States in the early 1950s by renowned architects such as Louis I. Kahn, Frank Lloyd Wright, Eero Saarinen and I. M. Pei (Grube 2005). It was part of the shifts in industrial and commercial architecture, ongoing since the mid-nineteenth century. New materials, innovative processing methods and advances in technology have successively facilitated more flexible spatial and structural solutions.

In the United States, the Second World War was the primary driving force behind the development of industrial architecture and the design of "Big Science" laboratories (Galison and Jones 1999, 498). Even before the war was over, the categorical distinction between structural engineering and architectural design ceased to apply: hybrid companies came into being and forged their reputation with technical innovations such as wide-span constructions, highly resistant load-bearing floors, and air conditioning and lighting systems. After peace was declared in 1945, businesses that had first made their mark with factory buildings—the Austin Company, for example—turned their attention to large-scale

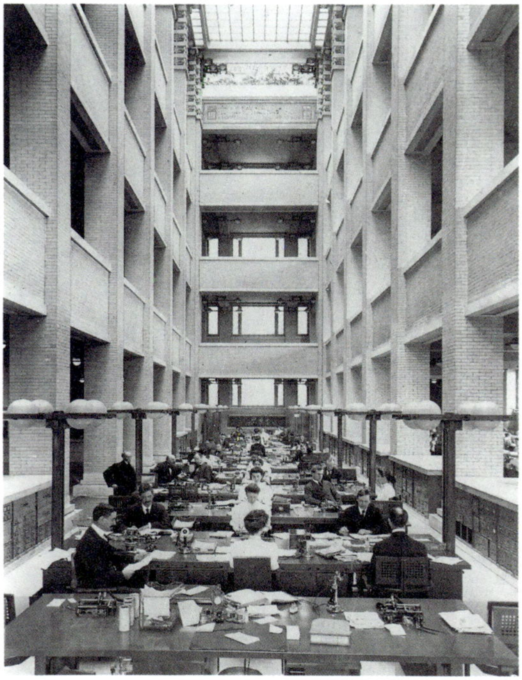

Fig. 5.2:
Frank Lloyd Wright, *Central top-lit court*, Larkin Company Administration Building, Buffalo, New York, 1902-06.
Photo © Courtesy of Buffalo History Museum, used by permission.

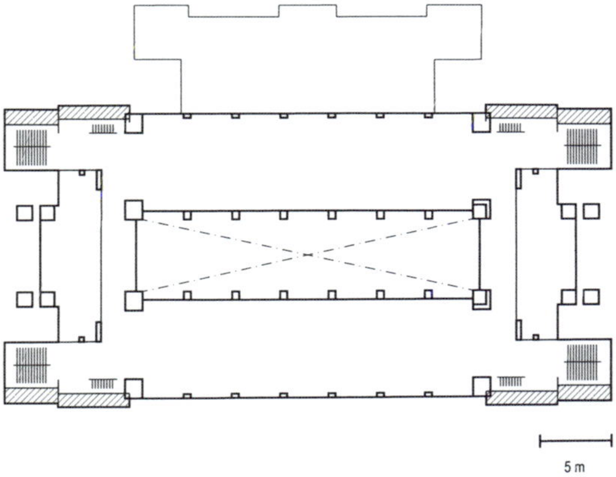

Fig. 5.3:
Typical floor plan, Larkin Company Administration Building.
Drawing © Patryk Kujawa.

laboratory buildings. While, prior to the war, laboratories with fixed walls had been the norm, flexibility now became a major design objective (idem, 498–99).

Office and industrial architectures likewise influenced one another. The work of Frank Lloyd Wright and other avant-garde architects pioneered the shift from the closed space to the flexibly configured open-plan zone. As early as 1906, Wright completed an office building composed of a central open space circumscribed by mezzanines, the Larkin Company Administration Building in Buffalo, New York (**Figs. 5.2** and **5.3**). This kind of well-tempered, continuous, open-plan space equipped with in-floor heating and air-conditioning presaged the office space so common today (Ingersoll 2008, 259).[1] According to Wright, the Larkin Building was a springboard for his further experimentation with open-plan designs:

> I think I first *consciously* began to try to beat the box in the Larkin Building—1904. I found a natural opening to the liberation I sought when (after a great struggle) I finally pushed the staircase towers out from the corners of the main building, made them into free-standing, individual features. Then the thing began to come through as you may see.
> (Wright and Kaufmann 1974, 284)

For Wright, freeing the corner from load-bearing elements, which he achieved primarily by shifting the supports, marked a decisive step "out of the box" and towards the open plan. With the SC Johnson Administration Building, completed in 1939, the architect

1 The administration building of the world's first mail order company—which epitomised the "panoptic machine" as defined by the French philosopher Michel Foucault (Foucault 1994)—was equipped with recreational and educational facilities for its employees. A light-flooded, clean interior space with an elegant dining room, roof terrace and conservatory as well as an organ for musical interludes and evening programmes was to assure a pleasant working atmosphere. Such important design features must now be regarded as forerunners of the trend in New Economy contemporary buildings towards more worker-oriented environments. Wright's Larkin Company Administration Building was a prototype built on principles that remain valid to this day. It was demolished in the early 1950s (Wright 1987, 126-27; Frampton 1994, 65-67).

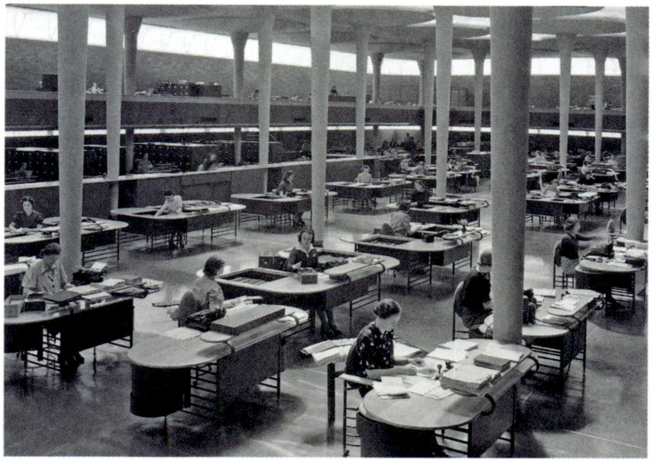

Fig. 5.4:
Frank Lloyd Wright, *Great Workroom*, SC Johnson Administration Building, Racine, Wisconsin, 1936-39. Photo © SC Johnson.

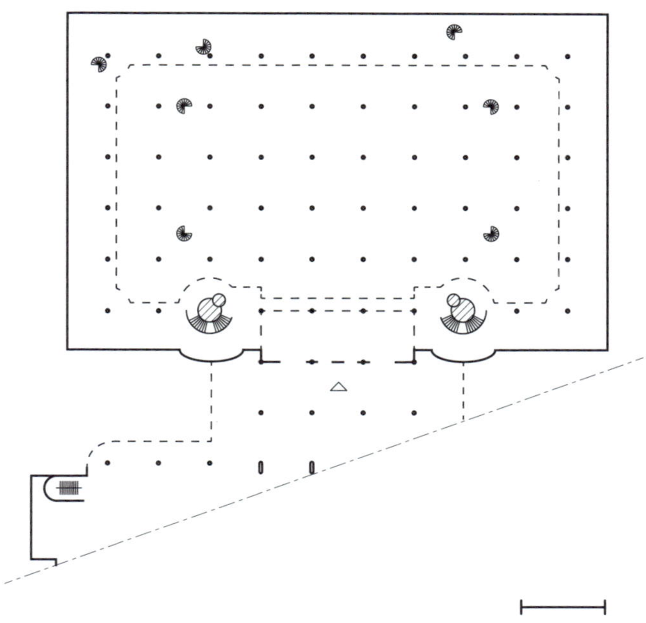

Fig. 5.5:
Ground floor plan, SC Johnson Administration Building. Drawing © Patryk Kujawa.

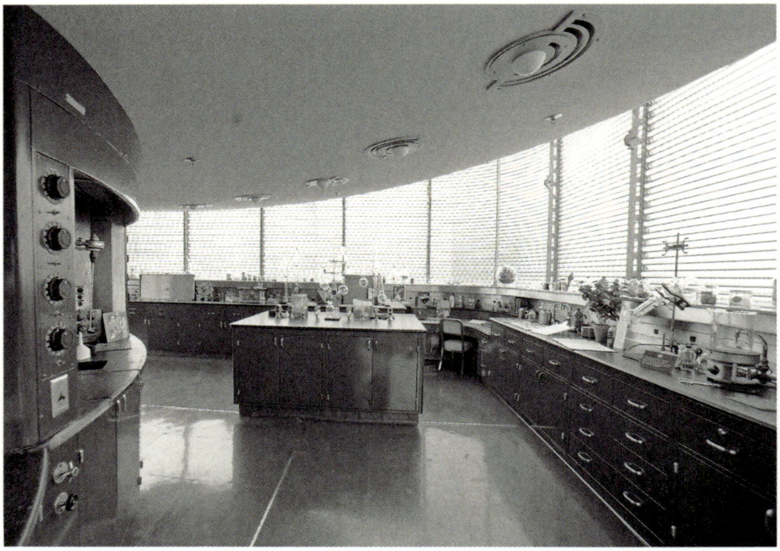

Fig. 5.6:
Frank Lloyd Wright, *Laboratory space*, SC Johnson Research Tower, Racine, Wisconsin, 1943-50. Photo © SC Johnson.

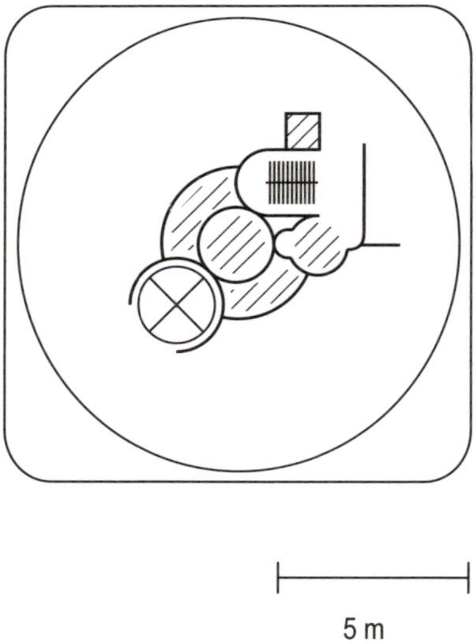

Fig. 5.7:
Typical floor plan, SC Johnson Research Tower. Drawing © Patryk Kujawa.

succeeded in creating an even bolder impression of continuous space, one that appeared to reach to the heavens due to the partially glazed roof (idem, 285–86) (**Figs. 5.4** and **5.5**). Wright finally transferred these spatial parameters to laboratory architecture with his SC Johnson Research Tower (1943–50) (**Figs. 5.6** and **5.7**):

> So we went up in the air around a giant central stack with floors branching from it, having clear light and space all around each floor. All laboratory space was then clear and in direct connection with a duct-system cast in the hollow reinforced-concrete floors, connecting to the vertical hollow of the stack itself. This seemed to me a natural solution [...] affording all kinds of delightful, sunlit, directly related workspace. (idem, 291)

The floor areas were small, however, and so the seemingly airy building actually lacked the flexibility required for certain scientific experiments (Alofsin 1994, 37).

The Salk Institute designed by Louis I. Kahn in the mid-1960s can therefore be regarded as the first building in which concepts such as optimal flexibility and adaptability were put into practice (Watch 2001, 1). A Vierendeel girder system[2] in the interim storeys facilitated wide spans and hence pillar-free, loft-style research spaces with a substantial degree of spatial and infrastructural flexibility. Yet Kahn's spatial achievements were hardly ever used. The interior design of the laboratory fell to an in-house planner who integrated a traditional laboratory layout in the open-plan spaces. Not a single spacious storey was left entirely open and the potential for flexible fittings was thus overlooked (Leslie 2005, 145–54).

To this day, the open-plan space with its inherent qualities of mutability and adaptability is usually fitted out conventionally. This is especially the case in biology or chemistry laboratories. For, while lab interiors are designed in part to meet specific needs

2 Named after its Belgian inventor, the Vierendeel girder system comprises structured frames with rectangular rather than triangulated members as well as fixed joints capable of transferring and resisting bending moments. Its open in-between spaces can be used, inter alia, for conduit wiring.

voiced by individual research teams, certain infrastructural and spatial components of any lab environment are subject to current international regulations. It is only recently that certain laboratories, particularly those engaged in interdisciplinary research, have begun to experiment with tailor-made solutions. Yet, until now at least, this too amounts to little more than a compilation of the

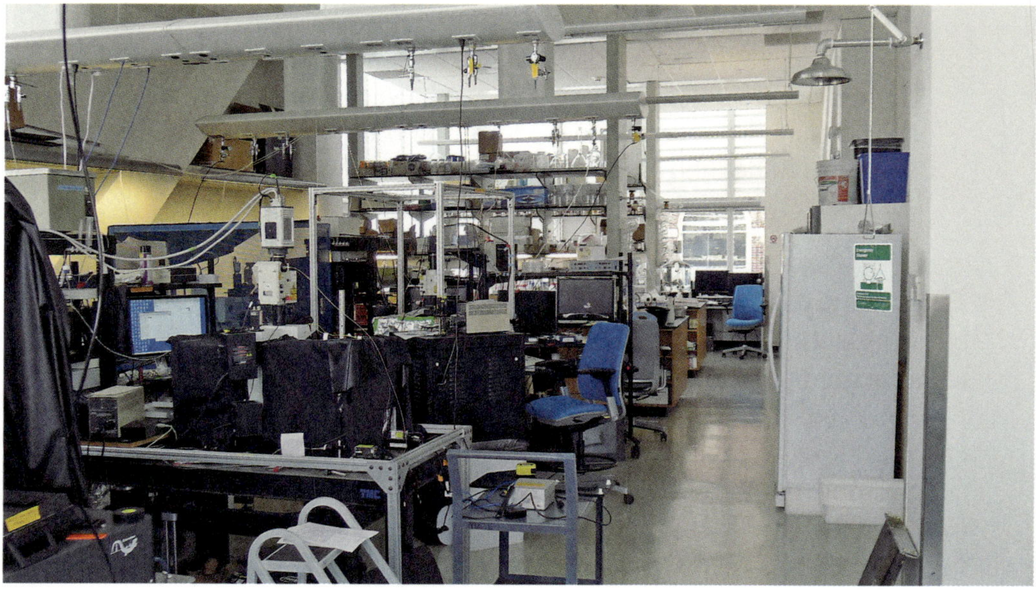

Fig. 5.8:
Rafael Moneo, Davis Brody Bond, and Moneo Brock Studio, *Laboratory interior with equipment for physics experiments in the foreground and conventional workbenches in the background*, Northwest Corner Building, Columbia University, New York, 2010. Photo © Sabine Hansmann.

different types of laboratory common to specific fields. Examples include the interdisciplinary labs in the Northwest Corner Building at Columbia University, New York. There, open-plan spaces are configured so as to meet the respective needs of different research teams. Rows of conventional chemistry workbenches, for example, alternate here with open spaces for physics experiments (**Fig. 5.8**). Although new laboratory buildings attest a clear trend to continuous large laboratory zones, serial installation of the generic, standardised, individual laboratory continues to be the norm.

The historian of science Thomas F. Gieryn used the Cornell Biotechnology Building (1988) in a case study to demonstrate which decisive factors ultimately made the standardised laboratory a common feature of the planning process. First and foremost among them was the decision to allocate the same size laboratory and office space to all scientists engaged at Cornell, regardless of their position or the number of students under their tutelage. "A cookie-cutter," Gieryn wrote, "could have been used to carve out the arrangement of wet benches, desks, sinks, free wall and floor space for heavy equipment, gas jets, and electrical outlets in every one of the generic labs in the building" (Gieryn 2002, 55). A combination of economic factors and higher education policy tipped the balance. For Davis, Brody & Associates, the architects of the building, the matter was simple: planning and seeking approval for only one standard module cost a lot less time and energy than negotiating countless individual preferences. Also, staying within budget was the major priority for both the administration at Cornell and the in-house project supervisors. To purchase standardised equipment or fittings in large quantities considerably lowers costs and also simplifies the matter of future care and maintenance. Moreover, making expensive customised furniture for faculty members just does not pay given that research staff tend to have limited tenure and thus may quickly move on. A standardised module has another advantage too: its uniformity fosters a sense of community that customised designs might easily undermine—or so it was presumed at Cornell (idem, 55–58).

The need for prestigious, signature architecture with which people can easily identify therefore conflicts with the need to design labs as flexibly and cost-effectively as possible. The planner's task consists in resolving the tension between the demand both for high-tech, functional and serviceable architecture and for a design that simultaneously assures a pleasant ambiance in the workplace and hence fosters communication. It is this "ambiance" that makes a building unique and puts a spin on the typically generic character of a laboratory building. As the art historian Ruth Hanisch points out, it is only by "slowly depositing the ambiance of a company or a product" that a "brand" is forged with which the distinctiveness of a building and the work pursued within it are henceforth associated (Hanisch 2011, 9–11). Yet exactly what impact a specific architectonic brand may have in the context of laboratory buildings is highly debatable. The need for flexible, functional laboratories is uncontested but the strategies pursued in order to lend this typology a special and specific air remain disparate.

ICONIC LABORATORY ENVELOPES

The "Bilbao effect" showed architecture's potential to reinforce a location's standing and usher in structural change.[3] Endeavours to use the physical reality of a building to upgrade a city, a company or a research institution are often undertaken in alliance with internationally renowned architects. The issue in such cases is no longer simply to design a building of the highest possible quality but rather to secure for the task a prominent architect who serves as a personal guarantee of quality.

For instance, the pharmaceutical company Novartis AG decided in 2001 to transform its base in the St. Johann district of Basel into a "campus of knowledge"—and to employ to this end star architects exclusively (See Chapter 4). The intention was to focus on people and their well-being, as Daniel Vasella, then CEO and Chairman of the company explained:

> It was becoming increasingly clear to us that the physical environment had a very substantial significance for the people there. An attractive work environment would also have a positive impact on our ability to recruit the best new talent. (Vasella 2008, 8)

3 Collaboration between the world famous cultural institution, the Guggenheim Foundation, and the star architect Frank O. Gehry on the construction of a branch of the Guggenheim Museum in Bilbao in northern Spain wrought a dramatic shift in a city hitherto shaped by its industrial past but now touted internationally as a new tourism hotspot. After opening in 1997, quasi as the flagship of a long-term development plan for the region, the new museum also triggered an economic upswing. The term "Bilbao effect" has since become a synonym for the use of iconic architecture to upgrade a location's profile.

Designed by renowned architects such as David Chipperfield, Tadao Ando, Frank O. Gehry or Rafael Moneo, every building on the new campus was presumed to be a potential architectural icon that would assure the company symbolic capital (Ingersoll 2008, 264).

The Salk Institute in La Jolla, California has long since become an architectural icon. The view across its inner courtyard is probably one of the most frequently published research centre motifs ever (**Fig. 2.1** [See Chapter 2]). Here, Louis I. Kahn pursued the interplay of meticulously specified volumes and the open inner space they framed. In addition, Kahn provided immutable spaces as a source of inspiration: "They should be sheer inspiration [...] just the place to be, the place which does not change, except for the people who go in and out. It is the kind of place that you enter many times" (quoted in Deubzer 2005, 18). In the Salk Institute these inspirational spaces consist in small private studies as well as in the large central inner courtyard. Both afford a highly rated view of the Pacific Ocean.

Kahn closely correlated internal functional requirements and exterior configurations. In his very first laboratory building, the Richards Medical Research Building at the University of Pennsylvania in Philadelphia (1957–61), he opened up the interior space and freed it of service shafts. He relocated service areas and utilities, both vertically and horizontally, in strikingly positioned brick towers (**Fig. 5.9**). The façade design took a backseat to functional components yet in a way so as to accentuate its iconic appearance. With this striking innovation Kahn lent new momentum to post-war discourse on the appropriate interaction of tradition and technology. Kahn developed his unusual design in collaboration with the mechanical engineer Fred Dublin and the engineer August Komendant, a pioneer in the use of pre-stressed reinforced concrete. The composition of monumental, weighty volumes gave expression to the building's inner structure and function. Technical and structural details were to remain as far as possible visible. In contrast to the elongated, loft-like laboratories Kahn created at the Salk Institute, the labs and offices at Richards are square modules, stacked one above the other in three seven-storey towers set around a central circulation and

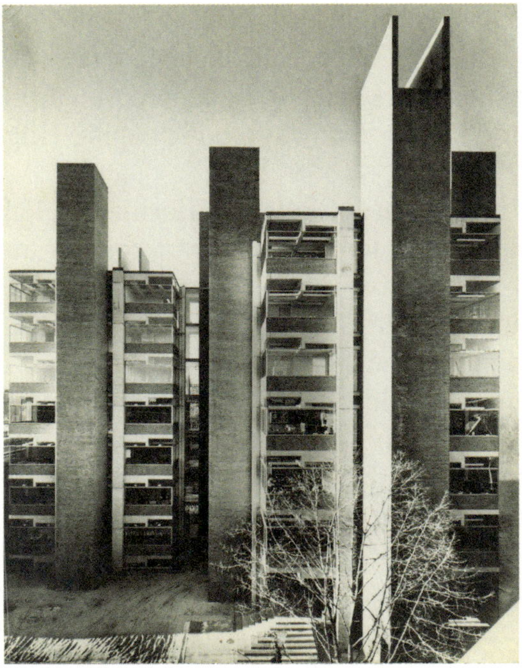

Fig. 5.9:
Louis I. Kahn, *View of the laboratory towers*, Richards Medical Research Laboratory, University of Pennsylvania, Philadelphia, 1957-60. Photo © Louis I. Kahn Collection, University of Pennsylvania and the Pennsylvania Historical and Museum Commission. Photo by Malcolm Smith.

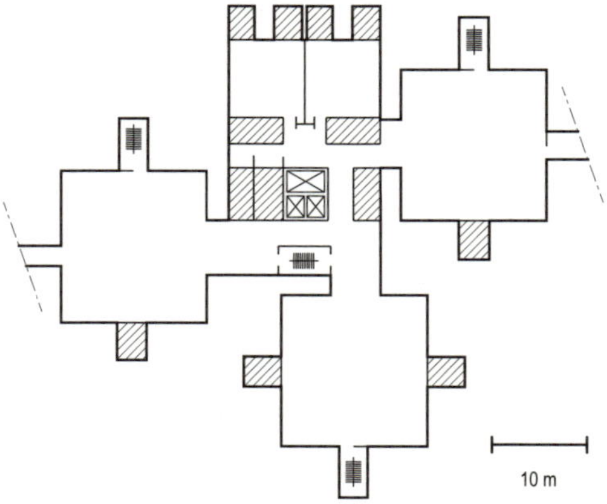

Fig. 5.10:
Typical floor plan, Richards Medical Research Laboratory. Drawing © Patryk Kujawa.

utilities tower (**Fig. 5.10**). Komendant developed for the building a modular system comprised of prefab concrete elements, which could be hoisted into place like a steel skeleton and then combined with in-situ concrete infill. Vierendeel trusses obviated the need for further supports and enabled the entire infrastructure to be accommodated in the load-bearing floor, which unlike at the Salk Institute was to remain unfaced (Leslie 2005, 91–127).

In 1961, the Museum of Modern Art declared the building to be "possibly the most significant example of post-war architecture in the United States" (The Museum of Modern Art 1961). Yet Venturi came to see it over thirty years later as a negative model, the epitome of everything a generic academic laboratory building should not be: "It is not an architectural vehicle for sculptural articulation that makes for expressionistic architecture that is heroic and original—that constricts flexibility and promotes distraction" (Venturi 1999, 395). Indeed, since the laboratories in the Richards Medical Research Building were relatively small and the brick towers ultimately accommodated only a part of the utilities infrastructure, the correlation here of function and aesthetics is somewhat ambiguous. Yet Kahn's relocation of the service shafts to the exterior walls—a move motivated by his refusal to install conduit wiring and pipes and thereby ruin his building (Banham 1984, 249)—was nonetheless a watershed in American architecture and shaped its development for many years to come (Grube 2005, 21).

Venturi deplored the extravagant architectonic gesture in the context of a laboratory building but was not at all averse to the necessity of a specific signature or symbolic charge. In particular the façade of a building lends itself to this latter purpose. And it is indeed often the case that renowned architects are commissioned to design the envelope as well as other conspicuous aspects of a research building whereas the design and realisation of the functional interior are left to experienced laboratory planners. The Lewis Thomas Laboratory designed by Venturi, Rauch & Scott Brown (as of 1989 Venturi, Scott Brown & Associates) is a prime and early example of such task segregation. Completed in 1986, it clearly shows that Princeton University was already in pursuit of strategic objectives:

> One of the primary objectives of Princeton University in constructing the Lewis Thomas Laboratory was to lure a team of the field's most prominent research scientists to the school. Ideally, the university hoped to achieve its goal by offering the potential faculty a state-of-the-art research and teaching laboratory. The new laboratory would challenge Princeton University's conservative construction policy by promising to be one of the largest buildings on campus, and undoubtedly one of the most expensive.
> (Collins 1999, 399–400)

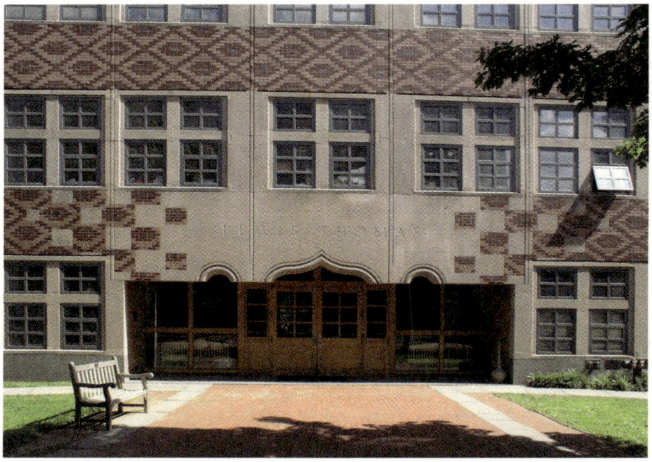

Fig. 5.11:
Venturi, Rauch & Scott Brown and Payette Associates, *Main entrance*, Lewis Thomas Laboratory, Princeton University, Princeton, 1983-86. Photo © Nicolas Janberg (www.structurae.de).

The architects were commissioned to plan the site and to design the building's envelope and entrance area while the laboratory specialists Payette Associates took on the design of the laboratories (**Figs. 5.11** and **5.12**). The choice of a generic laboratory typology within the building seemed appropriate given that eventual demands on it were largely hypothetical in the planning period, yet had already to be taken into account (idem, 405–08). The architectural project was above all an opportunity for Princeton University to consolidate its profile and make its molecular biology faculty more attractive to leading international research scientists (Gieryn 1999, 440). Whenever aspirations arise to use a laboratory building in this way, as a means to market a location, planners are required to come up with a specific architectonic position that transcends mere optimal functionality.

Even though laboratories are universally subject to standardisation, the specific configuration of a laboratory building and the emphasis placed on whatever is unique or characteristic seem increasingly to be a major focus. Venturi too complemented his personal paradigm—the requisite flexibility of laboratory buildings—by introducing features designed to instil a sense both of

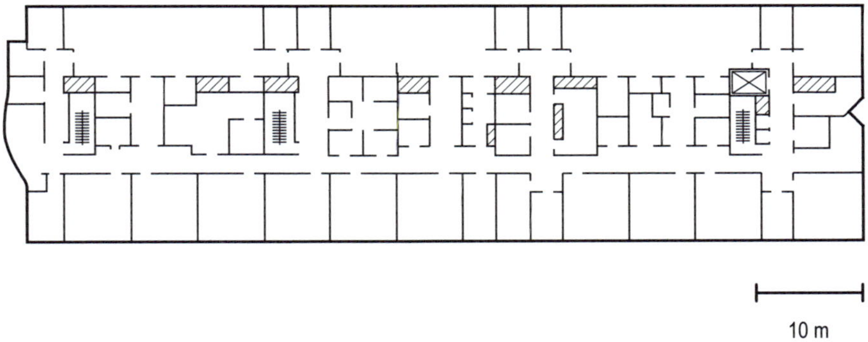

Fig. 5.12:
Typical floor plan, Lewis Thomas Laboratory.
Drawing © Patryk Kujawa.

community and of the specific location. His definition of the generic encompassed not only "functional ambiguity"—i.e. the need in his view for spaces able to be assigned multiple or changing purposes—but also the two poles "setting" and "place". "Setting", wrote Venturi, describes the laboratory in the broader sense, as the most neutral and reclusive "context space" possible, one that allows an individual or group to concentrate on the work at hand; whereas "place", by contrast, describes a space of communication that facilitates chance encounters and hence must remain as ambiguous as possible. "Place" brings into focus "the bigger picture", the scientific community as a whole, and therefore holds very particular significance: “Imageful Place [sic] that is local might be more essential than ever in our era of electronic communication to and from all over, in our era of networking” (Venturi 1999, 393). Venturi also distinguished between such internal requirements and outward appearances:

> The difference between outside and inside—at least in the lab building, and especially for the academic lab building—is very relevant where architectural consistency and neutrality of the workplace inside is counterbalanced by explicit symbolic content on the outside that acknowledges the significance of the institution as a whole.
> (idem)

Neutral interior space is to be counterbalanced thus by an explicitly symbolic context that may comprise ornaments, signs and iconographic references. Inasmuch as it constitutes a kind of rhythmic exception to the generic, this aspect, which Venturi called “symbolism & ornament”, generates aesthetic tension (idem). Here, the laboratory’s envelope, “likened to an Elizabethan manor house and an old schoolhouse” (Gieryn 1999, 450), quickly becomes a decorative marketing object that seeks to communicate with the general public first and foremost through its façade while simultaneously limiting public access to its interior, which is the preserve of the scientific community.

Hopkins Architects elaborated an altogether different type of envelope when commissioned to design an addition to the Frick Chemistry Laboratory on the Princeton University campus, completed in 2011 and located not far from the Lewis Thomas Laboratory. Here, the pivotal architectural feature is a public atrium (**Fig. 5.13**), a communicative space for impromptu encounters that perpetuates an impression of exterior space within the building. The envelope hence appears to have been turned inside out. The representative role of the building is likewise transposed from the exterior façade to the interior space, since the glazed atrium ensures that the work of the research scientists within remains largely visible. Like Norman Foster and Nicolas Grimshaw, Thomas Hopkins ranked among the leading lights in high-tech architecture of the 1970s and 1980s in the United Kingdom. As Design Architect of the Frick Chemistry Laboratory, Hopkins Architects developed a clearly structured volume of steel, glass and granite and, in collaboration with Payette Associates, a sophis-

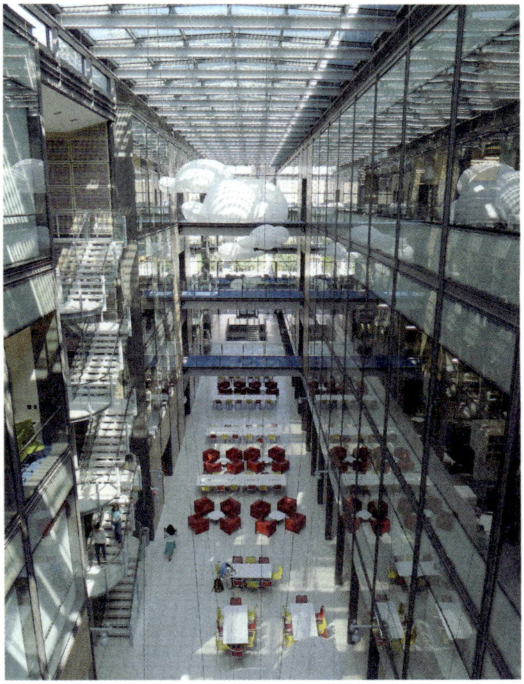

Fig. 5.13:
Hopkins Architects & Payette Associates, *Central atrium*, Frick Chemistry Laboratory, Princeton University, New York, 2007-10. Photo © Sabine Hansmann.

ticated building technology system with which the company hoped to set a new benchmark in energy-efficient design for chemistry research buildings in the United States (Gregory et al. 2012, 9). The building comprises a laboratory wing and an administration wing, with pedestrian bridges through the glazed atrium connecting the two. The front façade facing the campus renders this internal structure visible yet makes no direct reference to internal operations (**Fig. 5.14**). It is only up close that the ground floor demonstration and teaching labs come into view. Communication, both internal and public, is organised primarily via the atrium. The lab spaces as well as the offices have extensively glazed exterior walls, yet most of them give onto landscaped grounds somewhat removed from campus circulation routes. These façades therefore do not serve communication. The atrium is the linchpin and hub of the building, a dedicated space for interaction and circulation.

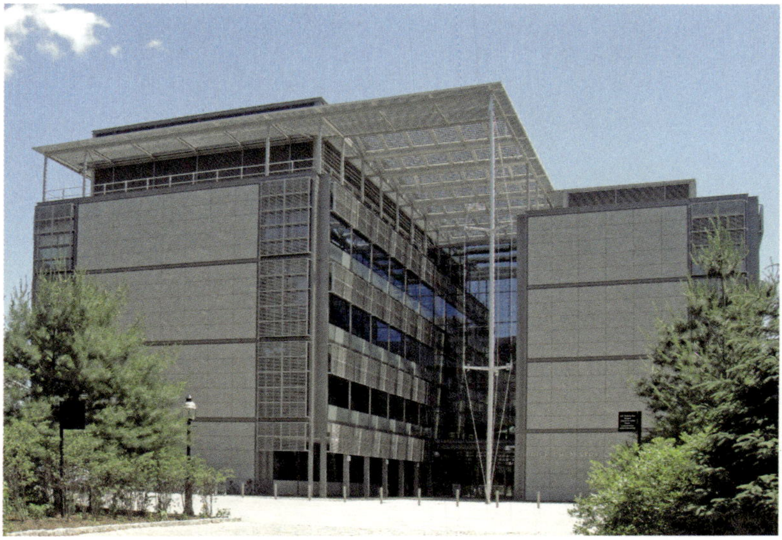

Fig. 5.14:
Elevation, Frick Chemistry Laboratory. Photo © Sabine Hansmann.

With its public café, study areas and various types of seating it offers ample opportunity for informal meetings (See Chapter 6). As deep as the entire building and over 20 meters high, it is also a stately space able to accommodate more formal events, such as university receptions. In everyday life it is a meeting place for staff as well as a window for the public onto the adjacent glazed laboratories and office tracts. In rendering visible the building's purpose, the internal façades put staff and their activities on display and thereby demystify them, at least in theory. Yet whether this ostensible transparency genuinely creates space for candid communication or is nothing more than a grand architectonic gesture remains an open question (See Chapter 4).

Thomas F. Gieryn identified a comparable instance of this "inside-out" approach—of a public zone able to assure maximum transparency within a building—in the James H. Clark Center completed in 2003 by Foster & Partners. While Robert E. Kohler describes the building as a prototypical "placeless place", decontextualised from its actual environment and equipped with uni-

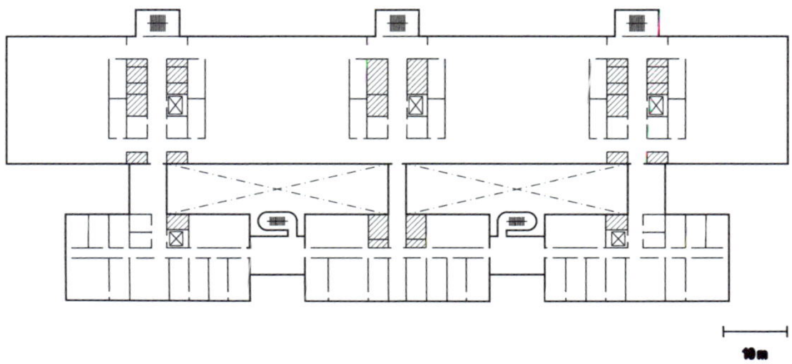

Fig. 5.15:
Typical floor plan, Frick Chemistry Laboratory. Drawing © Patryk Kujawa.

versal generic spaces (See Chapter 4), Gieryn, drawing on David Harvey and Manuel Castells, sees it as a building with specific features:

> First—and this is Harvey's point—hot buildings designed by branded architects become cargo cult landing strips with the potential to attract the wherewithal for successful but portable post-Fordist science. Second, Castells reminds us that creativity and innovation depend on: chance face-to-face interactions (and ensuing whiteboard or napkin outbursts); the mano-a-mano transmission of research skills and assumptions that do not travel well when formalized into words, numbers, or even pictures; agglomerations of rare talent, equipment, and research materials that must be immediately at hand for it all to work; and rapid-fire iterative adjustments in the face of failure or surprise. Virtual places won't do.
> (Gieryn 2008, 801)

In the Frick Chemistry Laboratory all three aforementioned aspects were realised. The building is largely independent of its context and its formal and functional design comprises large and uniform open labs as well as multi-functional office and consultation rooms. In this sense it seems to be a "placeless place" (Kohler 2008, 766). Yet the name and renown of the architects lend it iconic status. In Harvey's view, the simple fact that the commission went to star architects sufficed to assure the project a strong profile. To echo Castell, the laboratory's role as a space of interaction and encounter was arguably the more significant for Princeton University. Given the optimal transparency of its atrium, the Frick certainly is showcased as a place of encounter. The specific moment here, within this largely generic architecture, is assured solely by the visibility of the staff and their work.

CONCLUSION

Laboratory planning has given rise to international lab typologies, either of the open lab or lab module variety, which attest a high degree of standardisation and modularity. Laboratory design demonstrates advanced technological and infrastructural solutions for increasing requirements of generic flexibility and adaptability necessitated by the rapidly evolving network of global research. Yet large companies and institutions also make very specific demands: beyond purely functional criteria, their buildings must serve as an internationally marketable flagship and as a tool of internal and public communications. However, to create a distinctive ambiance that makes a location as appealing as possible to researchers, sponsors and partners and thereby to simultaneously provide a range of highly flexible and functional interior spaces poses a major challenge for architects. In contrast to other purpose-built, high-tech architectural typologies, for instance in the logistics sector, laboratory buildings do not appear at first glance to be generic. They attest rather—on account of their setting in a campus context, their relation to neighbouring buildings, their volumetric composition or the distinctive signature of their architects—a conscious endeavour to make a specific statement. And, as the following chapter demonstrates (See Chapter 6), the public zones and informal spaces within them are likewise increasingly being showcased. This trend is likely to gain ground in the future.

REFERENCES

Alofsin, Anthony. 1994. "Frank Lloyd Wright and Modernism." In *Frank Lloyd Wright: Architect*, edited by Terence Riley, 32-57. New York: Abrams.

Banham, Reyner. 1984. *The Architecture of the Well-Tempered Environment*. 2nd ed. London: Architectural Press.

Clynes, R. E. and A. J. Branton. 1984. "Flexibility and Adaptability—Alternative Design Strategies." In *Design, Construction, and Refurbishment of Laboratories*, edited by Ron Lees and Arthur F. Smith, 47-64. Chichester, UK: Ellis Horwood.

Collins, James. 1999. "The Design Process for the Human Workplace." In *The Architecture of Science*, edited by Peter Galison and Emily Thompson, 399-412. Cambridge, MA: MIT Press.

Deubzer, Hannelore. 2005. "Building Culture: Magic and Identity of Places." In *Research and Technology Buildings: A Design Manual*, edited by Hardo Braun and Dieter Grömling, 16-19. Basel: Birkhäuser.

Foucault, Michel. 1994. *Überwachen und Strafen: Die Geburt des Gefängnisses*. Frankfurt / Main: Suhrkamp.

Frampton, Kenneth. 1994. "Modernization and Mediation: Frank Lloyd Wright and the Impact of Technology." In *Frank Lloyd Wright: Architect*, edited by Terence Riley, 58-79. New York: Abrams.

Galison, Peter and Caroline A. Jones. 1999. "Factory, Laboratory, Studio: Dispersing Sites of Production." In *The Architecture of Science*, edited by Peter Galison and Emily Thompson, 497-540. Cambridge, MA.: MIT Press.

Gieryn, Thomas F. 1999. "Two Faces on Science: Building Identities for Molecular Biology and Biotechnology." In *The Architecture of Science*, edited by Peter Galison and Emily Thompson, 423-55. Cambridge, MA: MIT Press.

Gieryn, Thomas F. 2002. "What Buildings Do." *Theory and Society* 31 (1): 35-74.

Gieryn, Thomas F. 2008. "Laboratory Design for Post-Fordist Science." *Isis* 99 (4): 796-802.

Gregory, Rob, Paul Finch, and Michael Hopkins. 2012. *Hopkins 3*. New York: Prestel.

Grube, Oswald W. 2005. "Die Geburt des modernen Forschungsbaus in den USA." In *Entwurfsatlas Forschungs- und Technologiebau*, edited by Hardo Braun and Dieter Grömling, 21-27. Basel: Birkhäuser.

Hanisch, Ruth. 2011. "Atmosphären als Marketing?" In *Atmosphären*, edited by Markus Jager, 9-41. Dortmund: TU, Fak. Architektur, Städtebau und Bauingenieurwesen.

Hegger, Manfred. 2005. "Räumliche und technische Anforderungen an Forschungsbauten." In *Entwurfsatlas Forschungs- und Technologiebau*, edited by Hardo Braun and Dieter Grömling, 28-31. Basel: Birkhäuser.

Ingersoll, Richard. 2008. "From the Confinement of Heterotopia to the Urbanity of Novartis Ville." In *Novartis Campus: A Contemporary Work Environment: Premises, Elements, Perspectives*, edited by Jaqueline Burckhardt, Martine Francotte, Vittorio Magnago Lampugnani, Michael Renner, Wolfdietrich Schutz, 256-65. Ostfildern: Hatje Cantz.

Kohler, Robert E. 2002. *Landscapes and Labscapes: Exploring the Lab Field Border in Biology*. Chicago: University of Chicago Press.

Kohler, Robert E. 2008. "Lab History: Reflections." *Isis* 99 (4): 761-68.

Laboratory News. 2015. "Creating the Perfect Laboratory." <http://www.labnews.co.uk/features/creating-perfect-laboratory> (accessed 20 February 2015).

Leslie, Thomas. 2005. *Louis I. Kahn: Building Art, Building Science*. New York: George Braziller.

Stichweh, Rudolf. 1996. "Science in the System of World Society." *Social Science Information* 35 (2): 327-40.

The Museum of Modern Art. 1961. *Press Release, no. 56, 6 June 1961*. <http://www.moma.org/momaorg/shared/pdfs/docs/press_archives/2845/releases/MOMA_1961_0060_56.pdf?2010> (accessed 8 March 2015).

Vasella, Daniel. 2008. "Foreword". In *Novartis Campus: A Contemporary Work Environment: Premises, Elements, Perspectives*, edited by Jaqueline Burckhardt, Martine Francotte, Vittorio Magnago Lampugnani, Michael Renner, Wolfdietrich Schutz, 8-9. Ostfildern: Hatje Cantz.

Venturi, Robert. 1999. "Thoughts on the Architecture of the Scientific Workplace: Community, Change, and Continuity." In *The Architecture of Science*, edited by Peter Galison and Emily Thompson, 385-98. Cambridge, MA: MIT Press.

Watch, Daniel. 2001. *Building Type Basics for Research Laboratories*. New York: John Wiley.

Wentzlaff, Daniel, Maria Müller, and Ralf Streckwall. 2014. "Vitrinen der Innovation." *competition* 6 (1): 59-64.

Wright, Frank Lloyd. 1987. *Frank Lloyd Wright. Vol. 2. Monograph, 1902-1906*. Tokyo: ADAEdita.

Wright, Frank Lloyd and Edgar Kaufmann. 1974. *Writings and Buildings*. 3rd ed. New York: Nal Penguin.

“I consider that creating spaces for physical encounters is particularly important today, in order to foster personal exchange among research scientists.”

Norbert Koch, Department of Physics,
Humboldt University Berlin

6

INFORMAL SPACES IN LABORATORIES

Henrike Rabe

When commissioning the American architect I. M. Pei to design the Mesa Laboratory in Colorado in the early 1960s, the Directors, Walter Orr Roberts and Philip Thompson, wished for a place "where a variety of people can meet, privately or semi-privately, can be alone, or can be distracted by a different kind of beauty" (Roberts 1961, 2). They accordingly asked Pei "to avoid 'conventional patterns' of organizing the laboratory by discipline or by size, and […] urged him instead to encourage serendipitous encounters by including 'a place to pace' and 'nooks and crannies and irregular places where people can wander or sit and think'" (Leslie 2008, 188).

Roberts and Thompson's description spells out the assumption that knowledge production in a research context takes place not only in actual laboratories and offices but also at other sites within a building. Spaces not assigned a single purpose facilitate informal activities ranging from contemplative retreat to mutual exchange and interaction, from breaks to informal work-styles, and are equally relevant to scientific work. Such in-between spaces are described below as *informal spaces* in order to distinguish them from spaces of formal work (the laboratory or office).

In contemporary architectural discourse—some fifty-five years after Pei designed the Mesa Laboratory—terms such as "informal work", "communication", "interaction", "chance encounters", "serendipity", "social space", "breakout space" and "flexible learning area" are now ubiquitous in research, educational and office contexts (Bruffett 2015; Kojima and Akamatsu 2007;

Allen et al. 2004; Gieryn 2002). Especially in office planning, the relevance of *informal spaces* has become a matter of concern, in part owing to the research of Francis Duffy. Under the blanket term "club", they are attributed as important a role as any played by more formal types of space, such as the "cell" (individual office), the "den" (conference room) and the "hive" (open-plan office) (Duffy and Powell 1997). The architectural scholars Bill Hillier and Alan Penn highlighted the relationship between spatial structure and informal exchange in laboratory buildings as early as 1992 (Penn and Hillier 1992). In consequence, *informal spaces* have since gained in significance in that context too. Yet despite this, no or only limited provision is made for *informal spaces* on the whole, as recent architectural competition briefs attest—for example, those issued by the Berlin Senate Department for Urban Development in 2010 for the Forschungs- und Laborgebäude für Lebenswissenschaften (Research and Laboratory Centre for Life Sciences) and by the State Building Authority in Passau in 2014 for the Lehr- und Forschungsgebäude für Nachhaltige Chemie (Teaching and Research Centre for Sustainable Chemistry).

In this chapter, a selection of historical and more recent examples serves to elucidate how the ratio of informal to formal space has changed in research centres over the last few decades. Firstly, selected laboratory buildings are analysed in respect of the relative distribution therein of space and its respective function(s). Comparison of the floor area allotted to three types of space—here referred to as *spaces of theory, spaces of experiment* and *informal spaces*—reveals that the latter have gained considerably in significance since the 1990s. Secondly, the current design of *informal spaces* is analysed. Drawings are used to examine *informal spaces* in three recently completed buildings, taking into account their positions in the building, their dimensions, their openness, their fittings and furnishings, their ambiance and, finally, their usage.

AREA ANALYSIS IN TERMS OF FUNCTION

In order to examine shifts in the distribution of space in laboratory buildings, thirty-one buildings completed between 1960 and the present day were selected.[1] The 1960s laboratory projects of the architects Louis I. Kahn, Marcel Breuer and Sep Ruf were documented in various publications, so sufficient plans for their analysis are available. Well-documented buildings completed in subsequent decades by other renowned architects were likewise selected as well as a small number of laboratories to which access was assured by personal contacts. To guarantee comparability, four to six research buildings from different disciplines in the natural sciences or engineering fields were selected per decade, and two to three continents were taken into account. A minimum gross floor area of 1,000 square meters was a criterion for selection, since this paper focuses on laboratory buildings that serve a complex network of different functions.

Within the net internal area of the selected laboratory buildings, the following spatial functions could be identified: laboratory, collection depot, storage space, laboratory animal housing, office,

1 These laboratories are: the Max Planck Institute for Physics, Munich, DE (1960), the Chemistry Building of the Brookhaven National Laboratory, Upton, US (1960), the Richards Medical and David Goddard Laboratories, Philadelphia, US (1961), the Engineering Quadrangle, Princeton, US (1962), the Laboratorios de Física Avanzada of the Unidad Profesional Zacatenco, Mexico City, MX (1964), the Salk Institute for Biological Studies, La Jolla, US (1966), the Squibb & Sons World Headquarters, Princeton, US (1972), the Institute of Hygiene and Microbiology of the Free University, Berlin, DE (1974), the Medical Science Building of the University of Medicine and Dentistry of New Jersey, Newark, US (1975), Building S13 of the National University of Singapore, Singapore, SG (1977), the ARCO Chemical Company Research & Engineering Center, Newtown Square, US (1982), the Geology and Chemistry Laboratories of Brown University, Providence, US (1982), the Fraunhofer Institute for Building Physics, Stuttgart, DE (1983), the Lewis Thomas Laboratory, Princeton, US (1986), the Fraunhofer IPK, Berlin, DE (1986), the Max Planck Institute of Quantum Optics, Garching, DE (1986), the Bourns Hall Engineering Science Building, Riverside, US (1995), the Sir Alexander Fleming Building, London, UK (1998), the Fraunhofer IFAM, Bremen, DE (1999), the Amano Research Laboratories Gifu, Kagamigahara, JP (1999), the Biomedical Institute, Chongqing, CN (2001), the Donald Danforth Plant Science Center, St. Louis, US (2001), the James H. Clark Center, Stanford, US (2003), the Brain and Cognitive Science Center, Boston, US (2005), the NanoTecCenter, Weiz, AT (2007), the Novartis AG, Basel, CH (2010), the Northwest Corner Building, New York, US (2010), the Frick Chemistry Laboratory, Princeton, US (2010), the Sainsbury Laboratory, Cambridge, UK (2010), the UCSF Stem Cell Building, San Francisco, US (2011) and the ERIBA, Groningen, NE (2013).

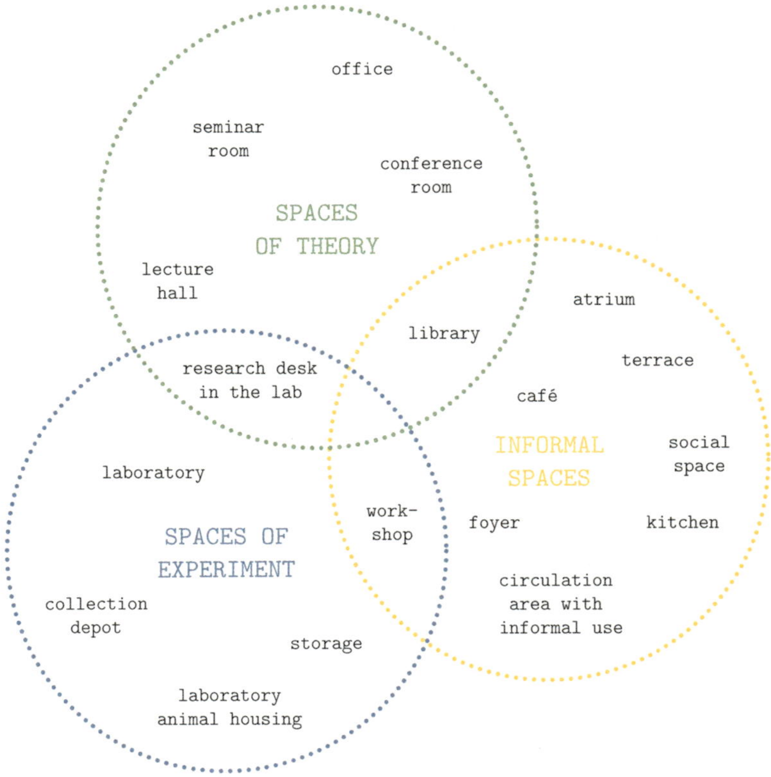

Fig. 6.1:
Categories of space according to function. Drawing © Henrike Rabe.

seminar room, conference room, lecture hall, atrium, foyer, café, social space, kitchen, library and terrace. Areas foreseen for informal uses within the circulation areas were likewise taken into account. Three categories of space were identified according to their function (**Fig. 6.1**):[2]

2 It is true that both informal activities and theoretical research can be pursued in all types of space, and experiments can be conducted beyond the borders of any single room or building (Galison and Jones 1999; Knorr Cetina 2001), but the main criterion by which spaces were categorised here was their primary function. Only by thus simplifying the matter was it possible to comparatively assess the different laboratory buildings.

1. *Spaces of experiment*, such as the laboratory, collection depot or laboratory animal housing
2. *Spaces of theory*, such as the office, seminar room, conference room or lecture hall
3. *Informal spaces*, such as the atrium, foyer, café, social space, kitchen, library, terrace or any area for informal use within the circulation areas

Comparison of the thirty-one selected laboratories in respect of the area distribution of the three functional categories *spaces of experiment*, *spaces of theory* and *informal spaces* enabled the following observations to be made: Until the 1980s, *informal spaces* occupied a relatively constant floor area, on average 8 per cent. This proportion started to slowly rise in the 1990s and currently averages 21 per cent (**Fig. 6.2**). Comparison of all thirty-one laboratories also demonstrates that the overall proportion of floor area assigned *informal spaces* in university laboratories is only 11 per cent, yet in commercial equivalents is significantly more, namely 18 per cent (**Fig. 6.3**). It further shows that the proportion of floor area

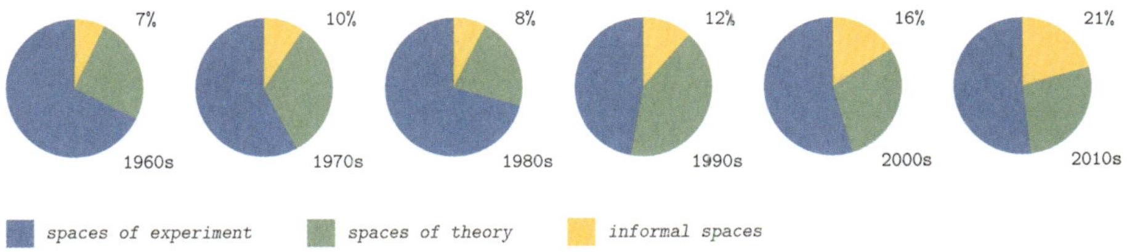

Fig. 6.2:
Area distribution in laboratory buildings in terms of completion date.

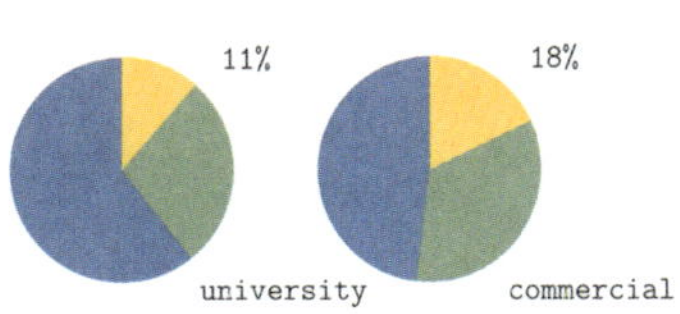

Fig. 6.3:
Area distribution in terms of laboratory type.

assigned *informal spaces* in university laboratories in Europe is 12 per cent and in North America 11 per cent, slightly more than in Asia (8 per cent) (**Fig. 6.4**).

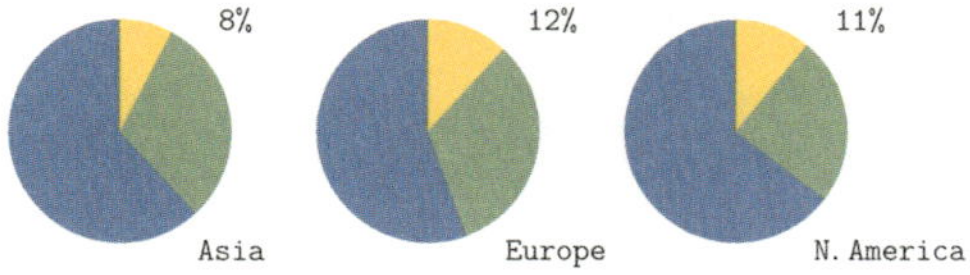

Fig. 6.4:
Area distribution in university laboratories in terms of continent.

Area analysis in terms of function clearly demonstrates that the floor area foreseen for *informal spaces* varies according to a laboratory building's construction date, type and continent. But it says nothing about how the increase since the 1990s in the amount of floor area foreseen for *informal spaces* affects the design of laboratory buildings.

SPATIAL ANALYSIS

In order to analyse the design of *informal spaces* in contemporary laboratory buildings, three more recent buildings were selected in which the floor area assigned *informal spaces* is especially extensive:

1. The Frick Chemistry Laboratory in Princeton
2. The ERIBA (European Research Institute for the Biology of Aging) in Groningen
3. The Northwest Corner Building in New York

The position of *informal spaces* in each laboratory building and their relationship to other spaces are to be clarified below and likewise whether they are open or closed, which types of work their fittings and furnishings facilitate, their ambiance and how their design is described and justified in texts about the buildings.

The analysis is based on published data, including drawings of and texts about the buildings, and complemented by personal observations, notes and sketches made in situ. Firstly, the floor plans of the buildings were analysed in order to identify the interrelation of the three spatial categories *spaces of theory, spaces of experiment* and *informal spaces.* Secondly, a so-called visibility graph analysis was made in order to ascertain the extent to which various points in a building are visible (Turner et al. 2001, 107).[3] And, finally, the potential movement of persons within the building was simulated with the aid of an "agent-based analysis" (Turner and Penn 2002, 473).[4] While this three-phase floor plan analysis first and foremost facilitated an understanding of the organisation of a typical floor, a further phase, namely analysis based on axonometric drawings, proved necessary in order to more closely examine the role of these *informal spaces* three-dimensionally. The diagrammatic character of all four phases combines analytical observation with a subjective reading of the building that is subsequently correlated with published architectural reports and other texts on the building.

3 Visibility graph analysis: with the aid of the software DepthmapX the number of points visible from any one standpoint in the building can be calculated. The findings are shown in the form of a heat map, whereby red stands for an especially high and blue for an especially low degree of visibility.

4 Agent-based analysis: the high frequentation of places such as elevators and stairways was integrated in this (likewise DepthmapX-facilitated) analysis by marking them as agents' starting points.

CASE STUDY 1: FRICK CHEMISTRY LABORATORY, PRINCETON, USA (22,296m²)

Hopkins Architects and Payette Associates designed the Frick Chemistry Laboratory, completed in 2010, on behalf of the Department of Chemistry at Princeton University (See Chapters 4 and 5). The freestanding building comprises two parallel, four-storey volumes connected by a street-style public atrium and several pedestrian bridges. *Spaces of experiment* are located in the larger laboratory wing to the northeast, and *spaces of theory* in the office wing to the southwest (**Fig. 6.5**). The former is divided into four sections and accommodates large open labs.[5] The latter has three sections interlinked by social spaces and open stairways. The layout of the two wings is identical on all floors except for the basement which accommodates an auditorium and additional laboratory space (Seward 2011).

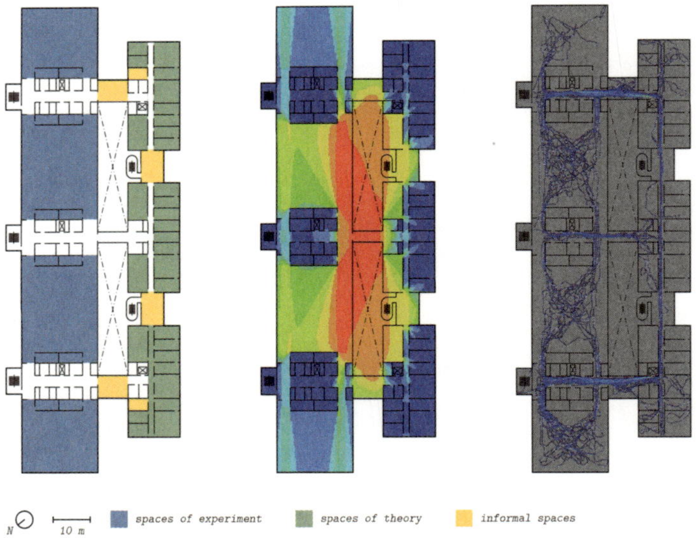

Figs. 6.5, 6.6 and 6.7:
Hopkins Architects & Payette Associates, *Typical floor plan*, Frick Chemistry Laboratory, Princeton University, New York, 2010. Left to right: functional analysis, visibility graph analysis, agent-based analysis. Drawings © Patryk Kujawa, Catherine Slusher.

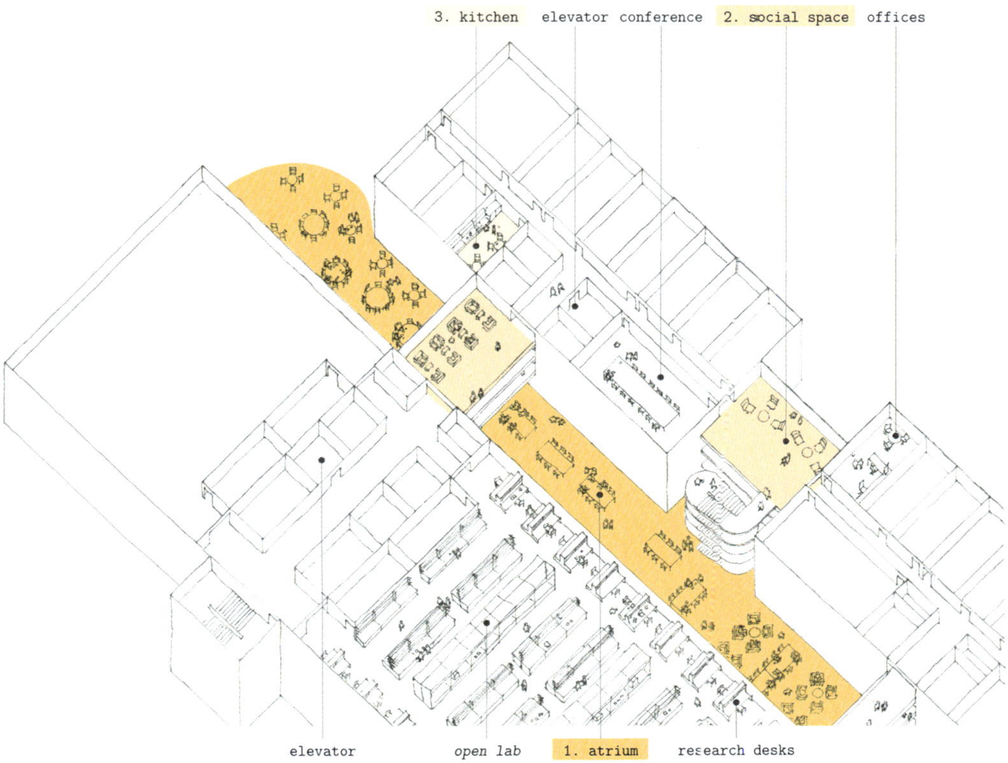

Fig. 6.8:
Axonometric analysis, Frick Chemistry Laboratory. Drawing © Henrike Rabe.

A total of three types of *informal spaces* can be identified **(Fig. 6.8)**:

1. The *atrium* is a central public street and meeting place comprising ground floor and basement spaces, pedestrian bridges at the upper levels and an outdoor terrace (**Figs. 5.13** and **6.9**). Extensive glazing assures a high degree of transparency as well as easy orientation throughout the building (**Fig. 6.6**). Fittings and furnishings such as study tables or lounge and

5 The term open lab—as opposed to individual or enclosed lab—describes a larger laboratory, often used by several research teams simultaneously (See Chapter 5).

café seating are intended to foster various activities (**Fig. 6.8**). Ceiling heights, transparency, natural light, the choice of materials (glass, steel, maple veneers) and brightly coloured furnishings combine to create an airy and inviting ambiance in this space, which is open also to the general public.

2. The spacious alcoves designated *social spaces* in the office wing are located near the open stairway to the atrium (Stephens 2010) (**Figs. 6.8** and **6.10**). As both the visibility graph and agent-based analyses demonstrate, they are set back from main circulation routes and therefore assure more privacy, in particular on the upper floors (**Figs. 6.6** and **6.7**).

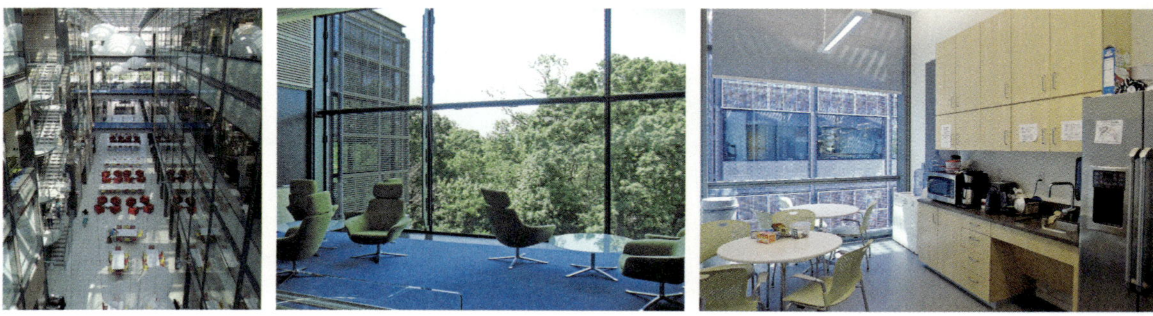

Figs. 6.9, 6.10 and 6.11:
Left to right: *atrium, social space, kitchen*, Frick Chemistry Laboratory.
Photos © Sabine Hansmann.

Armchairs, coffee tables, wall panels, kitchen sinks, drinking water dispensers and microwave ovens encourage both recreational breaks and informal styles of working. Soft furnishings, wall-to-wall carpets and the view of landscaped grounds beyond the floor-to-ceiling glazing further enhance the relaxing and, almost, homely ambiance of the *social spaces*.

3. Enclosed *kitchens* in the office wing (**Figs. 6.8** and **6.11**) are located close to the pedestrian bridges. As the agent-based analysis established, they are in frequent use (**Fig. 6.7**). Here too, a kitchen counter, small bistro tables and other fittings encourage breaks and informal styles of working, but in this case in a more sheltered spot (**Fig. 6.6**). The view of

the laboratories opposite enlivens the kitchens' functional ambiance.

Integrating *informal spaces* in the structure of this research centre was a major feature of the overall planning. As the architects' notes on their concept reveal, the intention was to design a building "which enhances interaction, collaboration and creativity" (Hopkins Architects 2011). Michael Hopkins explained that the atrium "[would] bring everyone together, help integrate general teaching and high-level research, and enhance collaboration and creativity" (Stephens 2010). Also, the "social spaces between the pods on the upper levels [were] intended as interaction zones for faculty and students" (idem). *Informal spaces* were therefore consciously situated at strategic points in the building, while their various orientations, degree of openness, furnishings and ambiance were designed to encourage a range of informal activities.

CASE STUDY 2: ERIBA, GRONINGEN, THE NETHERLANDS (10,400m²)

The ERIBA, a young yet pioneering collaboration between the University of Groningen and the University Medical Center Groningen, was designed by Rudy Uytenhaak Architectenbureau and opened in 2013. The typical upper floor of this seven-storey building is organised in two parallel strips around a slender central void. To the northeast are the *spaces of experiment*, to the

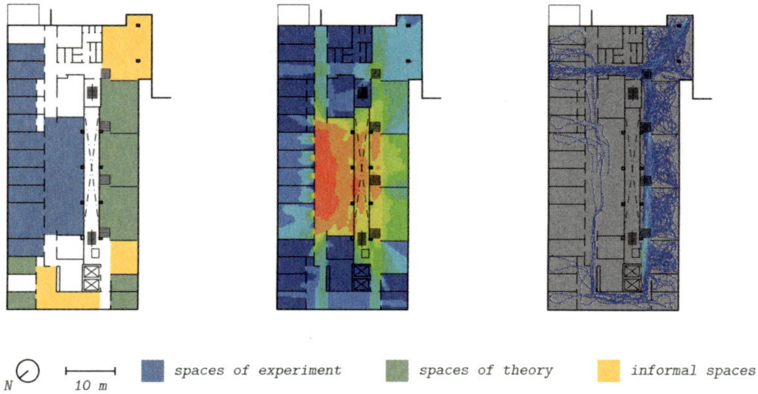

Figs. 6.12, 6.13 and 6.14:
Rudy Uytenhaak Architectenbureau, *First Floor*, ERIBA, Groningen, 2013. Left to right: functional analysis, visibility graph analysis, agent-based analysis. Drawings © Maren Krause, Catherine Slusher.

southwest the *spaces of theory* (**Fig. 6.12**). There is a lobby on the ground floor and a lecture hall on the third floor. The *spaces of experiment* comprise central open labs flanked by individual enclosed labs. The *spaces of theory* consist of large open-plan offices and auxiliary functions. Staggered floor heights on the first, second and third floors create distinct "plateau" areas within the open-plan offices and serve also as circulation routes.

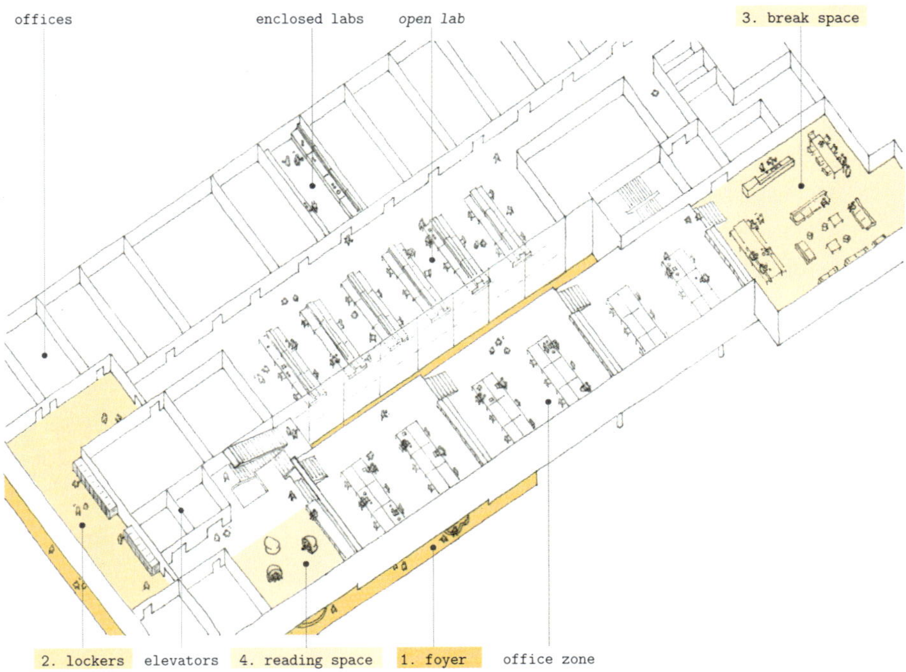

Fig. 6.15:
Axonometric analysis, ERIBA. Drawing © Henrike Rabe.

Four types of *informal space* can be identified at the ERIBA:

1. The public *foyer* is linked to the upper floors by a narrow continuous void containing an open stairway (**Fig. 6.16**). Extensive glazing affords views of the neighbouring laboratories and the grounds. A freestanding circular information desk, some exhibits and colourful amorphous seating arrangements structure distinct zones within the space. Given its location and design, the foyer serves mainly to communicate in-house research to the public rather than for actual research. Its connection with the upper floors, its furnishings and exhibits, and the view of the neighbouring laboratories are intended to make it an airy and inspiring space.
2. The *area with personal lockers* is close to the elevators (**Figs. 6.15** and **6.17**). As the visibility graph and agent-based analyses both demonstrate, this area between the laboratory

and office wings is heavily frequented yet largely hidden from view (**Figs. 6.13** and **6.14**). Lockers are a vital part of everyday working life here, since most employees are not assigned a fixed desk or workbench. Their location and function make an important hub and meeting place of this otherwise unfurnished and seemingly strictly utilitarian area.

3. Each of the *break spaces* opens onto the office wing (**Fig. 6.15**). Simulations demonstrate that these heavily frequented and highly visible spaces conveniently situated on the circulation route between the office and the laboratory wings are suitable not only for breaks but also "for small presentations and work discussions" (Permentier 2013, 3) (**Figs. 6.13** and **6.14**). Fittings, furnishings, materials and colour schemes vary from floor to floor in order to create different atmospheres. While, for example, armchairs, wooden tables and textile surfaces in warm colours are intended to give the first-floor break space a relaxed ambiance, its third-floor counterpart—the so-called "multifunctional" space—takes the form of a mezzanine overlooking the open-plan office below (Rudy Uytenhaak Architectenbureau 2010). Here, a bar and bar tables, the discreet use of colour and strikingly designed lamps, conjure a more formal atmosphere, making this an optimal space for special occasions as well as for more casual everyday encounters (**Fig. 6.18**).

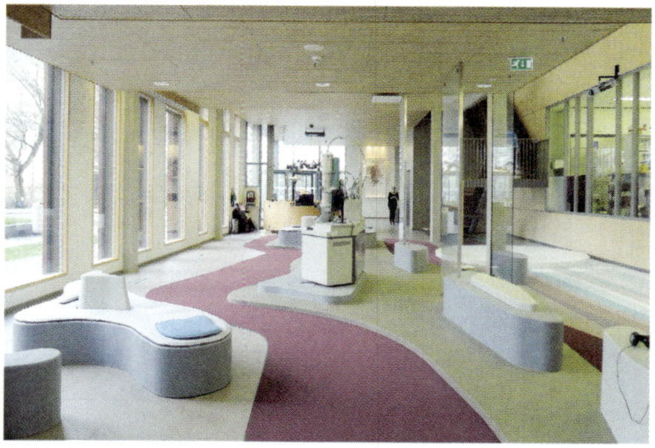

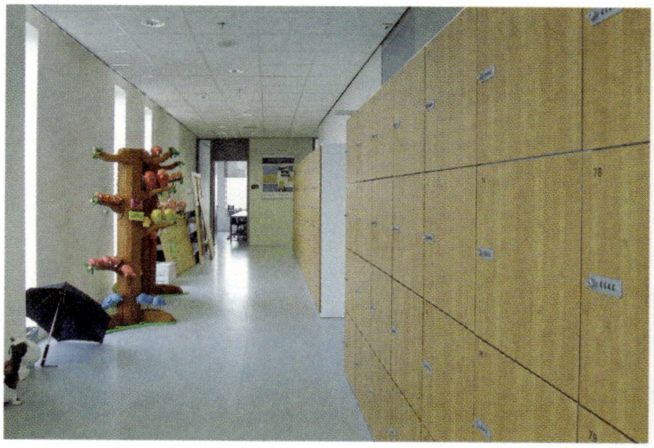

Figs. 6.16, 6.17, 6.18 and 6.19:
Left to right: *foyer, personal lockers, break space, pods*, ERIBA. Photos © Sabine Hansmann (6.16 and 6.17), Rudy Uytenhaak Architectenbureau (6.18), Friedrich Schmidgall (6.19).

4. Additional small *in-between spaces* are variously situated, designed and furnished. To name just one example, the cluster of pods by the entrance to the office wing is an invitation to do some reading or thinking (**Figs. 6.15** and **6.19**).

The ERIBA, with its four very distinctly designed types, offers a broad range of *informal spaces*. Descriptions of the project reveal how important it was to the clients that the architecture should foster interaction between research scientists, although *informal spaces* were not explicitly described as such: "Maximal transparency and visibility between the lab zones and the [...] work stations facilitates meetings and cooperation between the scientists, as wished" (Rudy Uytenhaak Architectenbureau 2010, 4). Planning for the office zones likewise took this point into account: "Given the paramount importance of mutual inspiration and of knowledge and data exchange, the work spaces have a particular form: a staggered arrangement inspired by terraced rice paddies provides the desired mix of seclusion and connectivity" (de Architect 2014). While at the Frick Chemistry Laboratory in Princeton interaction was emphasised above all else, at the ERIBA exchange and retreat are juxtaposed and accorded equal importance.

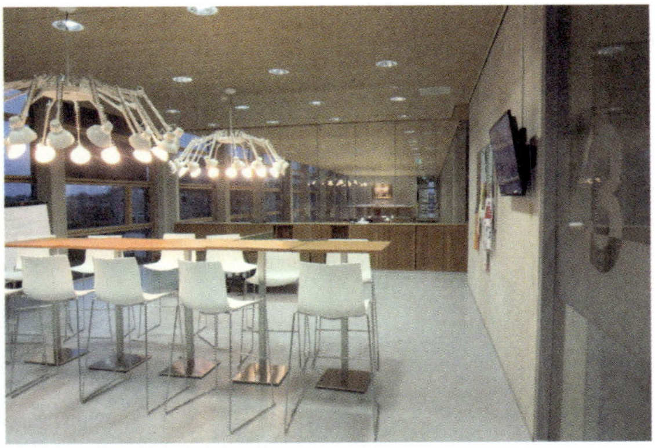

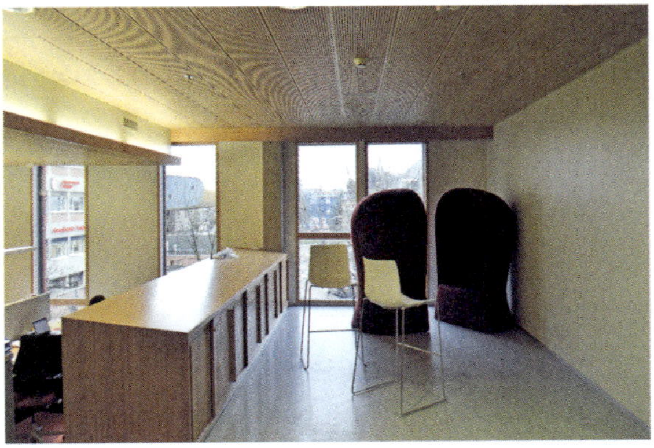

CASE STUDY 3: NORTHWEST CORNER BUILDING, NEW YORK, USA (17,468m²)

Rafael Moneo, Davis Brody Bond and Moneo Brock Studio completed the Northwest Corner Building at Columbia University in New York in 2010 (Seward 2009). The thirteen-storey building squeezed onto a corner lot above a pre-existing gymnasium hosts interdisciplinary research in various natural science disciplines. It is divided vertically into two sections. The six lower floors are structured by three large double-height spaces on different levels accommodating a café, a library and a lecture hall. The seven upper floors house the *spaces of experiment* and *spaces of theory.* To the northwest of the central corridor lie double-height open labs, to the southwest two-storey office zones (**Figs. 6.20** and **6.23**).

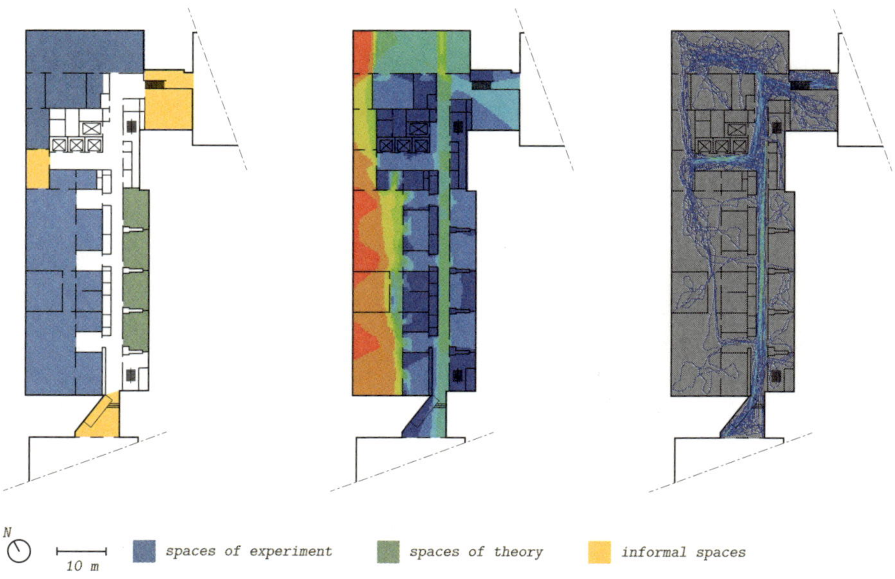

Figs. 6.20, 6.21 and 6.22:
Rafael Moneo, Davis Brody Bond & Moneo Brock Studio, *Seventh Floor*, Northwest Corner Building, New York, 2010. Left to right: functional analysis, visibility graph analysis, agent-based analysis. Drawings © Laura Enghusen, Catherine Slusher.

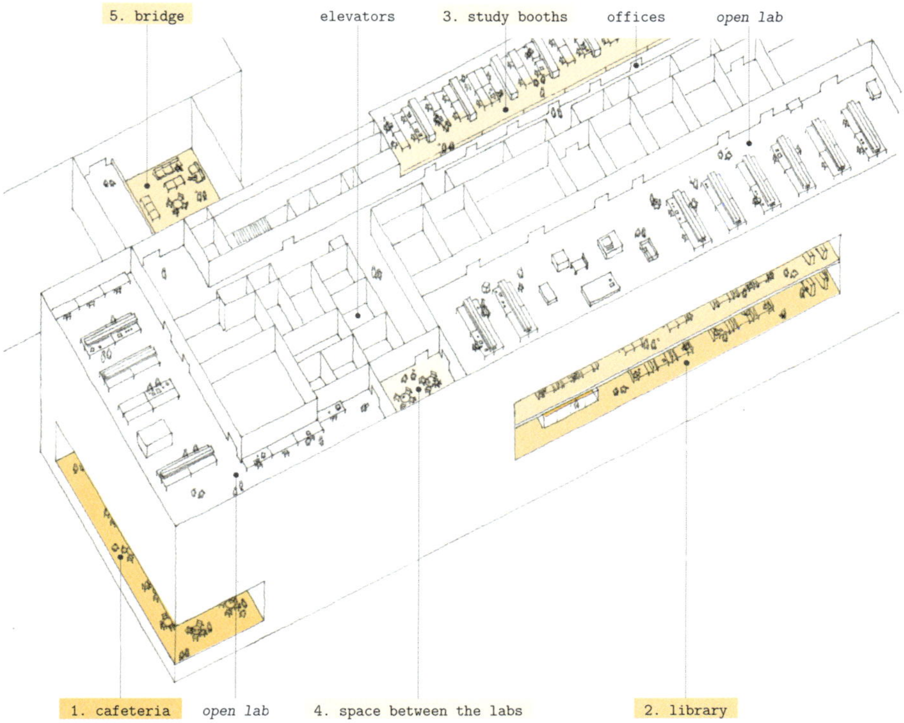

Fig. 6.23:
Axonometric analysis, Northwest Corner Building. Drawing © Henrike Rabe.

Here, five varieties of *informal space* can be distinguished.

1. The double-height glazed *cafeteria* close to the foyer is a station on the public route through the building (Sachare 2011, 4) (**Fig. 6.24**). Numerous bistro tables and a large bar are designed to foster informal activities. Generous dimensions, white marble walls and white furnishings further enhance the sleek and shiny impression made by this space.
2. The Science and Engineering *Library*, with its two reading rooms and several study rooms, extends from the fourth to the sixth floor (**Fig. 6.25**). "Shared work booths" and open tables in the reading rooms cater both to solitary study and to "students who wish to work together" (Landgraf 2012). Parquet flooring, louvered timber ceilings and timber furnishings create a warm and calming ambiance, even more so on the

low-ceilinged upper floor, where the warm glow of suspended light fittings makes a striking contrast to the functional look of the building's laboratories and offices.

3. The mezzanine on the laboratory floors accommodates students' *study booths* (**Figs. 6.23** and **6.26**) that open onto the unfurnished corridor and so facilitate spontaneous conversation between students working there and others passing by.
4. The small glazed *space between the laboratories* does not feature in the architects' published plans so was presumably built afterwards at the request of the users. It comprises two desks, a bistro table and kitchen facilities for the preparation of hot beverages and snacks (**Figs. 6.23** and **6.27**). As the visibility graph and agent-based analyses revealed, the striking features of the space are its transparency vis-à-vis the neighbouring laboratories and its accessibility from three sides (**Figs. 6.21** and **6.22**). The use of the same furnishings in this *informal space* as in the laboratories minimises differences in the ambiance of the two locations. Yet the former serves as a break space and interface for informal exchange between the two neighbouring workgroups.

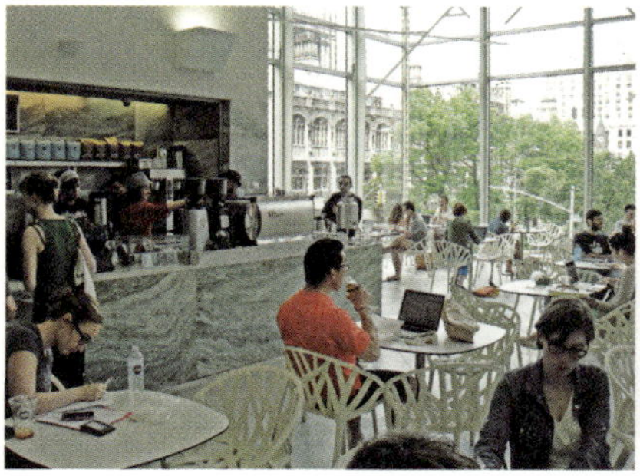
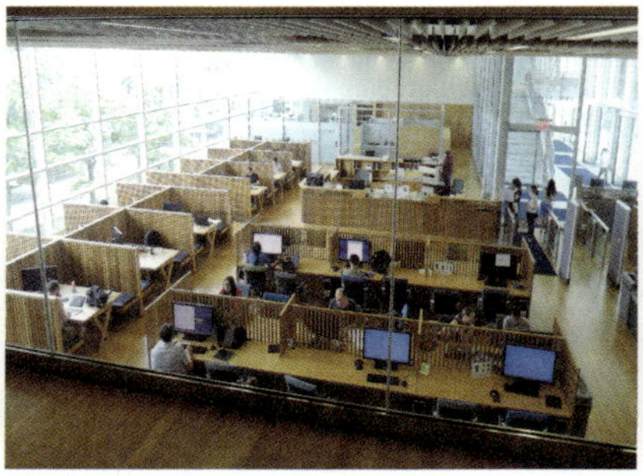
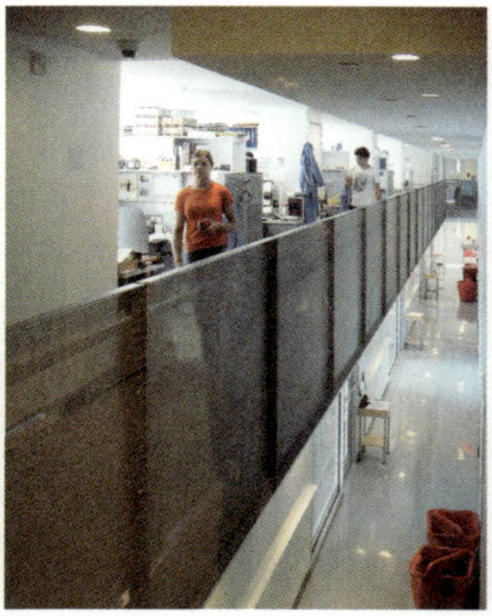

Figs. 6.24, 6.25, 6.26, 6.27 and 6.28:
Left to right: *cafeteria*, *library*, *study booths*, *space between laboratories*, *bridge*, Northwest Corner Building. Photos © Sabine Hansmann (6.25, 6.27, 6.28), Verena Straub (6.24, 6.26).

5. The location and design of further *informal spaces*, such as those of the *bridges* leading to neighbouring buildings and furnished with bistros and armchairs (**Fig. 6.28**), are likewise intended to foster spontaneous encounters, e.g. with scientists working in the adjacent departments.

Designed as an "interdisciplinary science building", the Northwest Corner Building "brings chemists, biologists, engineers, and physicists under one roof to facilitate the exchange of ideas and collaborations" (Columbia University 2013). It is also intended to "encourage interaction among faculty and students from the University's science and engineering departments" (idem). Its five different types of *informal space* serve different purposes and are located and designed accordingly. The *space between two laboratories*—evidently a late addition, since it can be discovered on a visit to the building but never featured in the plans—is proof that an easily accessible, adjacent break space for snacks and informal exchange is vital to anyone working in the laboratories and cannot be replaced by more distant *informal spaces*.

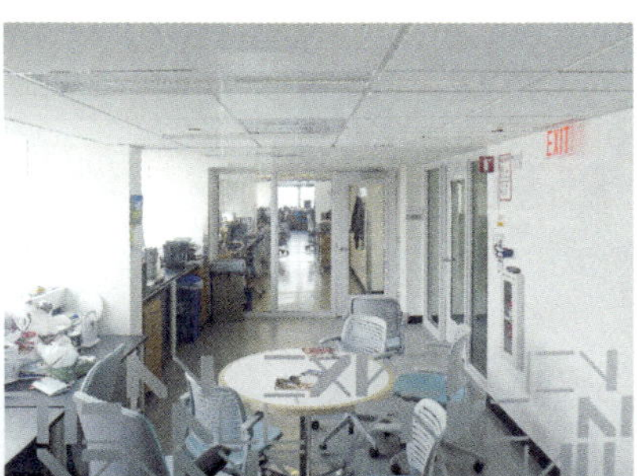

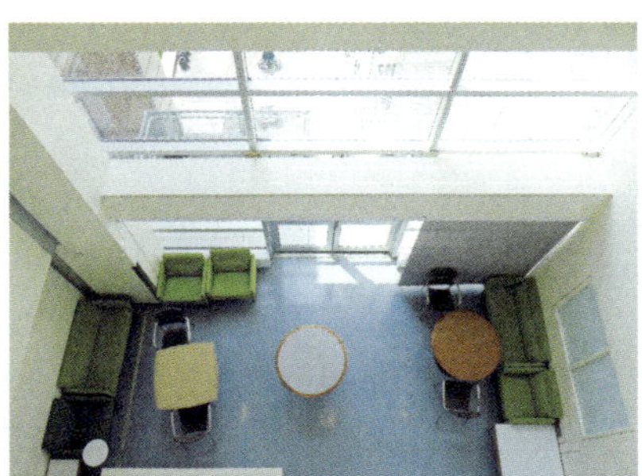

Analysis of the *informal spaces* in the three laboratory buildings shows how varied a form these may take. The differences can be categorised according to six parameters. Firstly, *informal spaces* are located at various strategic points in the buildings in accordance with their prospective usage: at central, easily accessible, often also publicly accessible points, or directly adjacent to laboratories or offices, or in more secluded spots. Secondly, the dimensions of *informal spaces* vary enormously, ranging from atriums as high as the respective building to double- or triple-height spaces to medium or small ones. Thirdly, various degrees of openness and transparency can be identified: whilst some spaces are fully exposed, others are open or transparent only on one or more sides, or are completely enclosed. Fourthly, different types of fittings and furnishings, such as study booths and conference tables, bistro chairs and armchairs, bar stools and bars, facilitate different types of activity. Fifthly, the use of varied materials and colour schemes conjures very different spatial effects and ambiances. Sixthly, a certain number of types of *informal spaces* can be found in all the examples studied, the greatest incremental value of which, one may assume, rests in their diversity and therefore in the diversity of uses they encourage. These usages—be they public or internal, quiet or communicative, or dedicated to work or to leisure—are united by the fact that they are not a formal component of scientific research but rather spontaneous, unscheduled or incidental.

The study of historical and more recent laboratory buildings demonstrates that so-called *informal spaces* have clearly gained in significance over time. The first step was to analyse the allocation of floor area to *spaces of theory, spaces of experiment* and *informal spaces* in selected laboratory buildings since 1960. This served to show that the area ratio of *informal spaces* has grown since the 1990s. The second step was to examine the *informal spaces* in three contemporary laboratory buildings. Six parameters were formulated for the analysis of their design: their position, their dimensions, their degree of openness, their fittings and furnishings, their ambiance and their diversity. The finding shows that informal spaces, thanks to their diverse designs and flexibility, are able to absorb various functions and types of usage and may therefore be of considerable importance in various situations in the everyday life of the research scientist. It therefore comes as no surprise that terms such as “interaction”, “communication”, “collaboration”, “creativity”, “connection” or “retreat” are common in the writings of architects and commissioning clients alike. In retrospect, it is clear that the requests that Walter Orr Roberts and Philip Thompson made of I. M. Pei, the architect of their MESA Laboratory, anticipated new trends in laboratory architecture.

REFERENCES

Allen, Tim, Adryan Bell, Richard Graham, Bridget Hardy, and Felicity Swaffer. 2004. *Working without Walls: An Insight into the Transforming Government Workplace*. London: Office of Government Commerce.

Bruffett, Marc. 2015. "Chance Encounters: The Winning Ticket to the Innovation Lottery?" <http://www.gensleron.com/work/2015/1/22/chance-encounters-the-winning-ticket-to-the-innovation-lotte.html> (accessed 19 July 2015).

Columbia University. 2013. "Chemists Expand into Northwest Corner Building." <http://www.columbia.edu/cu/chemistry/groups/newsletter/n1/newsletter1/nwc.html> (accessed 19 July 2015).

de Architect. 2014. "ERIBA, Groningen." Last modified August 27, 2014. <http://www.dearchitect.nl/projecten/2014/architectuur/eriba-groningen/eriba-groningen.html> (accessed 19 July 2015).

Duffy, Francis and Kenneth Powell. 1997. *The New Office*. 1st ed., London: Conran Octopus.

Galison, Peter and Caroline A. Jones. 1999. "Factory, Laboratory, Studio: Dispersing Sites of Production." In *The Architecture of Science*, edited by Peter Galison and Emily Thompson, 497-540. Cambridge, MA: MIT Press.

Gieryn, Thomas F. 2002. "What Buildings Do." *Theory and Society* 31 (1): 35-74.

Hopkins Architects. 2011. "Princeton Celebrates New Chemistry Building." <http://www.hopkins.co.uk/s/news/2011/111/> (accessed 19 July 2015).

Knorr Cetina, Karin. 2001. "Laboratory Studies: Historical Perspectives." In *International Encyclopedia of Social and Behavioral Sciences*, edited by Neil J. Smelser and Paul B. Baltes, 1st ed., 8232-38. Amsterdam: Pergamon.

Kojima, Kazuhiro, and Kazuko Akamatsu. 2007. *Cultivate CAt*. Tokyo: TOTO Shuppan.

Leslie, Stuart W. 2008. "‘A Different Kind of Beauty’: Scientific and Architectural Style in I. M. Pei’s Mesa Laboratory and Louis Kahn’s Salk Institute." *Historical Studies in the Natural Sciences* 38 (1): 173-221.

Landgraf, Greg. 2012. "Library Design Showcase 2012: Collaborative Learning." *American Libraries Magazine*, February 28. <http://americanlibrariesmagazine.org/2012/02/28/library-design-showcase-2012-collaborative-learning> (accessed 19 July 2015).

Penn, Alan and Bill Hillier. 1992. "The Social Potential of Buildings: Spatial Structure and the Innovative Milieu." In *Corporate Space and Architecture* (Proceedings of an international symposium, Lille, Lyons, Nantes, Paris, July 1-3 1992). <http://eprints.ucl.ac.uk/2705/1/2705.pdf> (accessed 19 July 2015).

Roberts, Walter Orr. 1961. "National Center for Atmospheric Research: Prospectus for a Laboratory." Boulder, CO: NCAR Archives. <http://nldr.library.ucar.edu/repository/collections/MLC-000-000-000-080> (accessed 19 July 2015).

Rudy Uytenhaak Architectenbureau. 2010. "ERIBA Groningen." <http://www.dearchitect.nl/binaries/content/assets/architect/projecten/2014/

architectuur/eriba-groningen/projectdocumentatie_eriba.pdf> (accessed 19 July 2015).

Sachare, Alex. 2011. "Northwest Corner Building Opens." *Columbia College Today* 38 (4): 4.

Seward, Aaron. 2009. "In Detail: Columbia U.'s Northwest Corner Building." <http://archpaper.com/news/articles.asp?id=3545#.VNoZZSke6-g> (accessed 19 July 2015).

Seward, Aaron. 2011. "In Detail: Frick Chemistry Laboratory. Hopkins Architects with Payette and MVVA Create a Machine in a Garden at Princeton University." <http://archpaper.com/news/articles.asp?id=5303#.VRaztim5Lnc> (accessed 19 July 2015).

Stephens, Ruth. 2010. "Elements of New Frick Lab Join to Create 'Best Infrastructure' for Chemistry." <http://www.princeton.edu/main/news/archive/S28/32/85K84/> (accessed 19 July 2015).

Turner, Alasdair, Maria Doxa, David O'Sullivan, and Alan Penn. 2001. "From Isovists to Visibility Graphs: A Methodology for the Analysis of Architectural Space." *Environment and Planning B: Planning and Design* 28 (1): 103-21.

Turner, Alasdair and Alan Penn. 2002. "Encoding Natural Movement as an Agent-Based System: An Investigation into Human Pedestrian Behaviour in the Built Environment." *Environment and Planning B: Planning and Design* 29 (4): 473-90.

BERLIN CASE STUDIES

“Differences in the design of lab spaces worldwide that I have noticed were institutional and temporal, not cultural, which is to say, they were shaped far more by the wealth and (private, government, public) nature of the institution than by the era in which the labs were built.”

Antoine Kahn, Department of Electrical Engineering, Princeton University, USA

7

THE INSTITUTE OF CHEMISTRY AT THE HUMBOLDT UNIVERSITY BERLIN FROM 1900 TO THE PRESENT

Ronald Göbel

In 1900 the chemist and future Nobel Laureate Emil Fischer launched operations at The First Institute of Chemistry at the Humboldt University Berlin (**Fig. 7.1**).[1] He had made this new laboratory on Hessische Strasse a prerequisite of his continued tenure at the University and regarded it furthermore as a new architectural benchmark for the design of teaching and research institutions in the natural sciences. As Fischer remarked in his inauguration speech, "no institute of this kind anywhere in the world can outshine the diversity and wealth of the means and methods of production [available here]" (Meisel 2011, 21). A number of renowned scientists were to work there in the following years, among them Walther Nernst and Fritz Haber, both Nobel Laureates like Fischer himself, as well as Otto Hahn and Lise Meitner. Yet apart from the international standing of the scientists engaged in research there, the building is of interest also with regard to the spatial organisation of its teaching and research. In this chapter the distribution of space in The First Institute will be examined and compared with later developments. How much floor area, it will be asked, is given to theoretical work, how much to practical work, and how does this relate to the thematic orientation of teaching and research activities.

1 The Humboldt University Berlin has borne this name only since 1949. Before that, it was called the Friedrich Wilhelm University. For simplicity's sake, the current name is used throughout this chapter.

In order to determine how much emphasis was placed on practical lab-based instruction, the spatial organisation of the Institute is considered below in the light of its curriculum, which can be reconstructed by recourse to the indexes of lectures published from 1900 to the present. The history of The First Institute of Chemistry serves thus as an exemplary case study of a research and teaching institution, the architectural development of which attests not only shifts in policy on scientific research but also the impact of broader political events. For example, damage suffered during the Second World War and the subsequent division of Germany left their mark not only on the curricula but also on the design of the laboratory building.

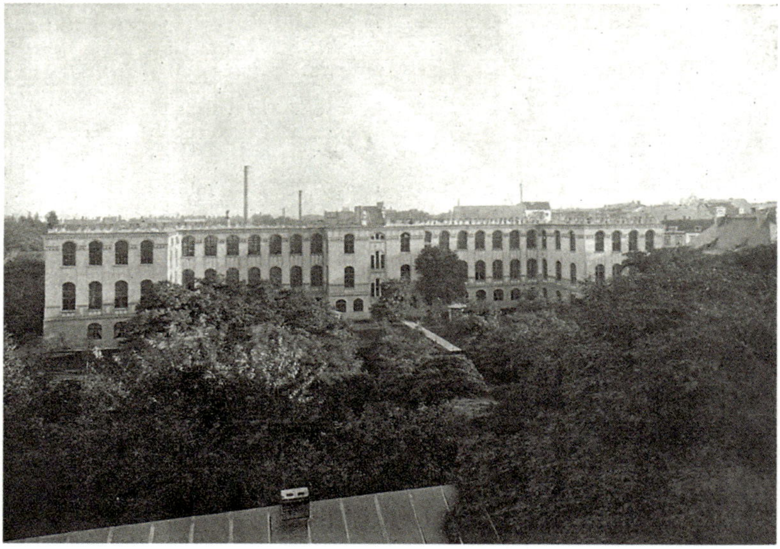

Fig. 7.1:
Northern elevation, The First Institute of Chemistry, from: Fischer and Guth 1901: book cover.

THE FIRST INSTITUTE OF CHEMISTRY ON HESSISCHE STRASSE FROM 1900 TO 1945

In 1892, following the death of August Wilhelm von Hofmann, Director of the Institute of Chemistry at the Humboldt University Berlin, the highly esteemed chemist Emil Fischer was appointed his successor. Fischer's acceptance of the post was contingent on the construction of a new laboratory building specifically designed to meet the demands of the day on research and teaching. Yet planning for the new building was repeatedly put on hold, and Fischer therefore threatened in 1896 to leave the Humboldt University in order to succeed August Kekulé at the University of

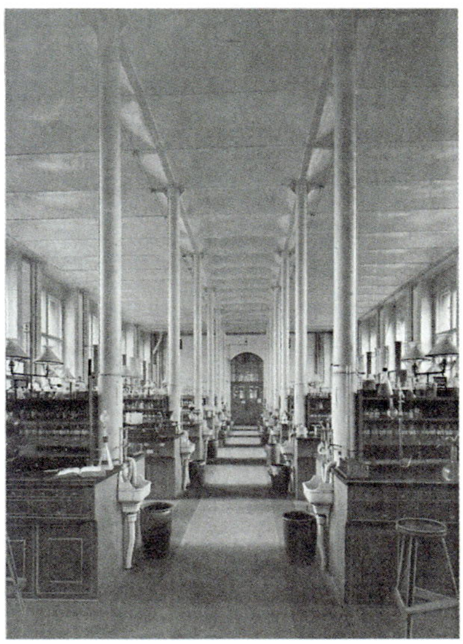

Fig. 7.2:
Teaching Laboratory I, The First Institute of Chemistry, from: Fischer and Guth 1901: 29.

Bonn. Thanks, however, both to Fischer's personal commitment and the support of the lobby group "Verein zur Wahrung der Interessen der chemischen Industrie" (Association for the Protection of the Interests of the Chemical Industry), construction of the new Institute began in 1897, on the site of the former Charité cemetery on Hessische Strasse (Meisel 2000, 11). The architects Georg Thür and Max Guth drew on Fischer's own proposals when designing the laboratory building, which was handed over to the Humboldt University on 14 July 1900 (Fischer 1900).

Structurally, The First Institute mirrored the educational ideal posited by Wilhelm von Humboldt when the University was founded in 1809, namely that research and teaching should be inseparable (Benner 2003). Conceived primarily for the practical instruction of a new generation of scientists, the building comprised at its core four large teaching labs that were flanked by research laboratories allotted to the various scientific departments (**Fig. 7.2**). Most of the space in the building was earmarked for lectures and lab-based practical work. The four large teaching labs—two each for the Departments of Organic and Inorganic Chemistry respectively—were situated on the first and second storeys of the so-called *Zwischenbauten* (intermediate sections) located to either side of the *Mittelbau* (central section) of the building (**Fig. 7.3**). Each lab had a floor area of circa 270 square meters and accommodated workbenches for circa 40 students. Thus circa 160 students were able to work simultaneously on a total floor area of approximately 1,080 square meters. Additional space was

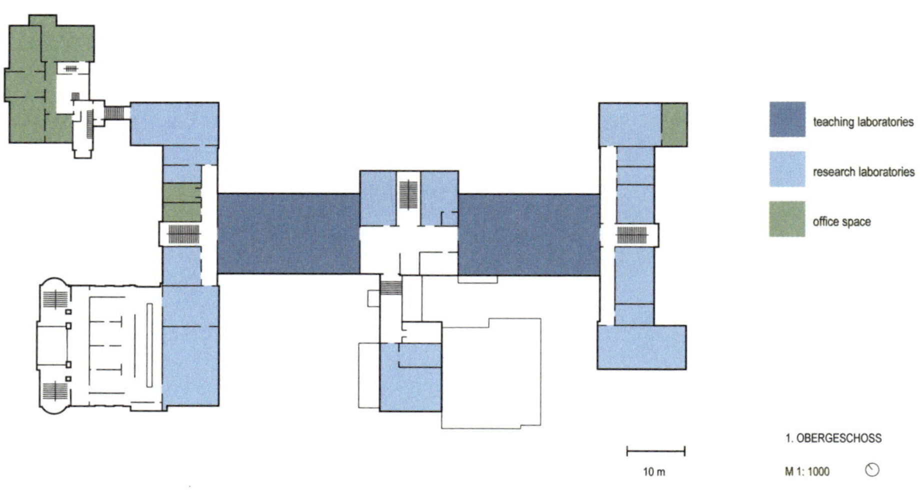

Fig. 7.3:
Ground plan of the 1st floor, The First Institute of Chemistry, from: Fischer and Guth 1901: Plate III.

available for practical work on special experiments. The research laboratories were located in lateral wings directly adjacent to the teaching labs, and visitors could traverse the latter to reach them.[2] The architecture of the building thus expressed the link between and equal standing of research and teaching.

	Inorganic Chemistry	Organic Chemistry	Physical Chemistry	Total
Research Lab	450	353	383	1,186
Teaching Lab	970	870	---	1,840
Total	1,420	1,223	383	3,026

Table 1:
Size of the laboratories (in m^2) at The First Institute of Chemistry, from: Fischer and Guth 1901: 63-65.

The floor areas listed in Table 1 show that the ratio of research lab to teaching lab was circa 1:1.5 (**Table 1**). The instruction of students in chemistry was clearly a primary focus in the building, and scientific research only secondary, at least in terms of the space allotted to it. A prestigious institution with leading scientists clearly did not preclude a strong commitment to tertiary education. Lectures were held in The First Institute's three large lecture halls, which could accommodate 634 students on a total floor area of circa 500 m^2. If one takes into account the floor area of the lecture halls, the ratio of space devoted overall to research respectively to teaching is 1:2.

2 The research laboratories, including, inter alia, private laboratories, weighing rooms and collection storage depots, had a total floor area of 1,186 square meters (Fischer and Guth 1901, 63-65).

THE DEVELOPMENT OF CHEMISTRY STUDIES AT THE HUMBOLDT UNIVERSITY FROM 1900 TO 1945

No mandatory curriculum for students of chemistry existed at the start of the twentieth century, and in fact formal curricula were introduced at the Humboldt University only in 1975. A direct comparison of the subjects and methods of study pursued in the interim is therefore not feasible. However, the number of hours taught and the types of teaching available in 1900, as recorded in the Humboldt University's annual index of staff and lectures, give some indication of the relevance of the various types of chemistry teaching then current (Friedrich-Wilhelms-Universität Berlin 1900). Theoretical instruction consisted in lectures, seminars and tutorials, and practical instruction in "hands-on" work in the laboratories. The following graphics show the number of hours taught (**Fig. 7.4**) and the range of courses offered (**Fig. 7.5**) at the Humboldt University.

In the academic year of 1900, the chemistry curriculum offered almost 70 different theoretical courses (consisting of lectures, seminars and tutorials) amounting in total to 150 hours per week, as well as 17 complementary practical courses amounting in total to 660 hours a week. These figures include the courses offered by affiliated scientific institutions, such as the Institute of Physical Chemistry on Bunsenstrasse (on the university campus in Berlin-Mitte) and the Kaiser Wilhelm Institute of Physical Chemistry and Electrochemistry in Berlin-Dahlem. A similarly diverse range of theoretical courses—almost 75 in number—was still available ten years later. However, the number of hours taught had shrunk by then by around 20 per cent, owing to a reduction in the average length of courses. In the same decade, the number of practical courses rose from 17 to 27, which represented, in comparison with the academic year 1900, a 60 per cent rise in the number of hours taught, to 1,060 hours in total.

This shift in curriculum focus, from theoretical instruction to practical "hands-on" instruction in the laboratory, suggests that the professional demands made on graduates had changed. As mentioned at the beginning of this chapter, the new institute on

Hessische Strasse came about thanks not only to Emil Fischer's personal commitment but also to the support of the chemical industry lobby (Meisel 2000, 11). In its striving for self-reliance and military supplies, the State of Prussia saw fit to found the *Kriegsrohstoffabteilung* (Raw Materials Department of the War Ministry), which was empowered to deem certain areas of research or even whole institutions vital to the war effort. This meant that scientists and engineers involved in relevant research were assured priority funding (Zigan 2007, 27). Industrial and military developments required chemistry teaching to diversify and to foster specialist expertise both in industrial chemistry and chemical engineering. An increase not only in the number but also in the duration of lab-based practical courses went some way towards meeting these new demands.

A further expansion of lab-based practical courses can be noted over the following decade. Thus the total time devoted to lab-based practical courses was raised by about 20 per cent in the academic year 1920, from 1,060 to 1,300 hours. By contrast, the total time devoted to theoretical courses dropped by almost 40 per cent, from 124 to 77 hours (Friedrich-Wilhelms-Universität Berlin 1920). Here too, the interests of the chemical industry lobby may explain the persistent focus on lab-based practical instruction. The period 1910 to 1920 saw the development of a whole string of commercial synthesis procedures for the inexpensive production of basic chemicals such as ammonia, sulphuric acid and nitric acid. The necessity of further refining industrial synthesis procedures explains the chemical industry's commitment to a lab-based, practical form of university education. It can also be surmised that personal acquaintance played a part in the revision and adaptation of curricula. The Humboldt University, for instance, offered practical courses led by Fritz Haber, who began working in 1909 with Carl Bosch, later Chairman of the Board at BASF, on what came to be known as the Haber process for the commercial synthesis of ammonia. Haber was appointed to the Humboldt University in 1916, initially as a private lecturer (Humboldt-Universität zu Berlin, Biographie Fritz Haber). He personally guaranteed that links were forged between the current needs of the chemical industry and the future of university education.

From 1900 to 1930, the number of courses and hours taught steadily developed without any major leaps. A slight drop in the amount of time devoted to lab-based practical instruction can be noted shortly after the outbreak of the Second World War: roughly 10 per cent, from 1,440 to 1,300 hours (Friedrich-Wilhelms-Universität Berlin 1940).[3] This trend continued after the end of the Second World War (Universität Berlin 1946).

THE FIRST INSTITUTE OF CHEMISTRY AND CURRICULA DEVELOPMENT AT THE HUMBOLDT UNIVERSITY AFTER 1945

Several air raids in the winter of 1945 severely damaged the Institute and put a temporary end to research and teaching. Coordinated reconstruction of the building began in 1946 under the supervision of the freshly appointed chemistry professor Erich Thilo. It was largely complete by late 1952 and everyday operations then resumed, although the official inauguration took place only in January 1953 (Thilo 1953). The design and dimensions of the original building were mainly retained (Herneck 1976, 212). The core feature was still four teaching labs for practical experiments, flanked by research laboratories. There was one essential change, however, regarding office space. While before 1945 no more than two office spaces had been available for consultations, the postwar period saw the gradual conversion of space hitherto reserved either for residential purposes or technical infrastructure. Thus the original Director's House, consisting of circa 500 square meters and adjoining the Institute, was now converted into offices.

But how did this spatial reshuffle correlate with the thematic orientation of chemistry studies at the Humboldt University? A

3 In both 1930 and 1940, 67 different lectures, seminars and tutorials were offered (following a temporary decline in the number of courses in 1920, to 52 different types of instruction), which more or less corresponded to the level in 1900 and 1910. The total number of hours taught per week accordingly rose to 117 (1930) respectively 96 (1940). The number of different practical courses offered in the same period remained almost constant, ranging only from 32 to 34.

glance at the index of lectures for 1950, four years after provisional labs had gone into service, shows that the number of theoretical courses on offer was only half that of the pre-war period and the number of hours were accordingly down by circa 40 per cent (**Fig. 7.4**). This state of affairs can be explained partly by wartime damage, which restricted use of the building, and partly by the dismissal of certain faculty members on political grounds. The number of hours devoted to practical courses, however, was surprisingly similar to that prior to wartime damage:

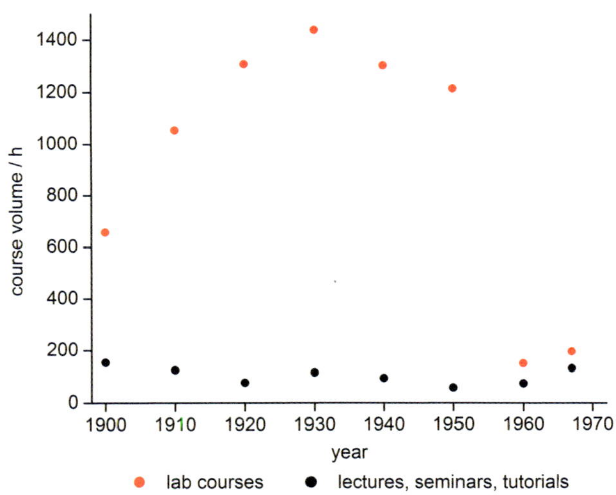

Fig. 7.4:
Number of hours taught in the Chemistry Department of the Humboldt University from 1900 to 1967, from: Friedrich-Wilhelms-Universität Berlin 1900-1940, and Universität Berlin / Humboldt-Universität zu Berlin 1946-67.

29 course options were available, only slightly less than in 1940 (Humboldt-Universität zu Berlin 1950). In those parts of the Institute that had suffered only limited damage or already been repaired, provisional laboratories for practical work had been set up and put into service in January 1946 (Thilo 1953). Regular operations resumed at the Institute in late 1952, upon completion of almost all the reconstruction work (Humboldt-Universität zu

Berlin 1952). While the goal of the latter had been to recreate as far as possible the Institute's exact original layout, the actual curriculum underwent extensive revision several times before 1960 (**Figs. 7.4** and **7.5**). The number both of theoretical courses and hours taught rose slightly, while the varied range of practical courses was reduced in number from 29 to 10 and now constituted only one-third of total course content. There was also a substantial drop in the number of hours devoted to lab-based work, namely from 1,200 to a mere 150 hours a week.

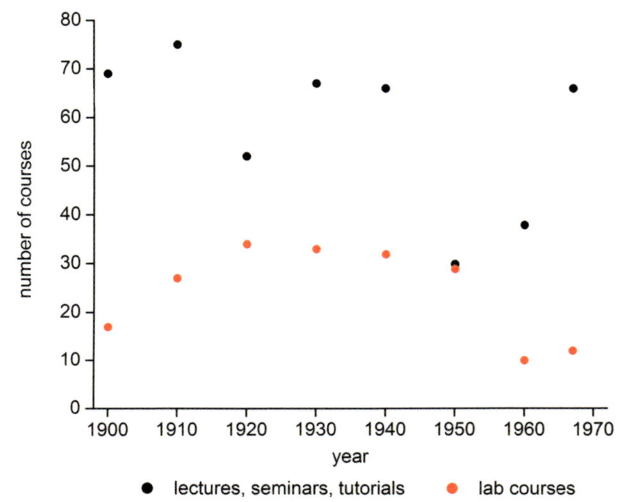

Fig. 7.5:
Number of courses taught in the Chemistry Department of the Humboldt University, according to type, from: Friedrich-Wilhelms-Universität Berlin, 1900-1940, and Universität Berlin / Humboldt-Universität zu Berlin 1946-67.

Since reconstruction of the Institute was by now almost complete, changes to the curriculum must be attributed to the University's revision of its educational policy. The drastic reduction in course options virtually robbed students of the chance to structure their studies individually and in accordance with their primary interests. The time devoted to practical instruction rose by 1967 to circa 200 hours per week yet failed by far to attain the pre-war level of circa 1,400 hours (Humboldt-Universität zu Berlin 1967). Since no annual index of lectures was published between 1968 and the year of German reunification, 1990, data for this period is derived from curricula published by the Ministry of Education in the GDR (Ministerium für Hoch- und Fachschulwesen 1975). The following table shows in hours the amount of time devoted to theoretical and practical instruction in chemistry studies in the period from 1975 to 2009 (**Table 2**).

Year	Lectures, Seminars, Tutorials	Practical Work	Total Hours	Time Ratio
1975*	1,050-1,500	1,000-1,500	≈ 2,500	1 : 1.5
1982*	1,050-1,500	1,000-1,500	≈ 2,500	1 : 1.5
1992	1,380	2,055	3,435	1 : 1.5
1997	1,470	1,425	2,895	1 : 1
2009**	1,680	1,430	3,110	1 : 0.9

Table 2:
Number of hours taught in chemistry studies at the Humboldt University from 1975 to 2009.
* Variations in each category depend on the area of specialisation. The total number of hours taught in each area of specialisation was circa 2,500.
** The statistics are for the MSc Chemistry course.

In 1975, depending on a student's specialisation, eleven to fifteen different lectures and seminars were foreseen, the total number of hours taught ranged from 1,050 to 1,200, and the six to eight practical courses on offer amounted in total to between 1,000 and 1,500 hours (idem). The ratio of time devoted to theoretical instruction respective to practical instruction was now circa 1:1.5 and thus attests an emphasis on lab-based practical work (**Fig. 7.6**). This curriculum was reapproved in 1982 and remained unchanged until 1992.

Upon German reunification in 1990, the Chemistry Department of the Humboldt University was disbanded and appointment proposals were invited for every teaching position (Meisel 2000, 19). The curriculum introduced in 1975 ceded in 1992 to a new study schedule accompanied by recommendations for a typical course of study. The number of hours taught increased by circa 30 per cent, to 1,380 hours. The number of practical courses offered rose by 40 per cent, to 2,050 hours. While the total number of hours increased by over one-third, the ratio of time spent on theoretical respectively to practical courses was 1:1.5, almost the same as in 1975 (Humboldt-Universität zu Berlin 1992).[4] This ratio changed only when course requirements were updated in the year 1997. The total number of hours taught (lectures, seminars

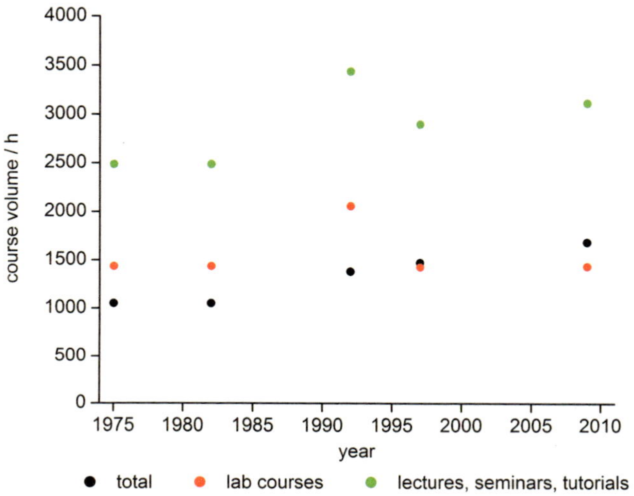

Fig. 7.6:
Time allocation in chemistry studies at the Humboldt University from 1975 to 2009, from: Ministerium für Hoch- und Fachschulwesen 1975 (1982) and Humboldt-Universität zu Berlin 1992, 1997 and 2009.

and tutorials) was raised by 7 per cent to 1,470 while lab-based practical work was reduced by circa 30 per cent to 1,425 hours. The number of hours devoted to theoretical instruction respectively to lab-based practical instruction was now at a ratio of 1:1

4 The increase in the overall number of hours required for an academic qualification, from circa 2,500 to 3,435 hours, was possible because the hitherto mandatory "political education" classes ceased. These have not hitherto been taken into account in this chapter, but amounted in total to circa 600 hours.

(Humboldt-Universität zu Berlin 1997). For the first time ever, these two components of chemistry studies carried equal weight (**Fig. 7.6**).

A further shift in course content occurred when Bachelor and Master courses were introduced in Germany. When the curriculum was revised in 2009, the time devoted to theoretical courses rose by more than 10 per cent to 1,680 hours, while the time devoted to lab-based practical instruction remained virtually unchanged. This marked a turnaround: now, for the first time ever, more time was allotted theoretical instruction than lab-based practical instruction (Humboldt-Universität zu Berlin 2009).

PLANS FOR THE NEW INSTITUTE IN BERLIN-ADLERSHOF

Upon reunification in 1990, Berlin suddenly found it had three universities with chemistry departments, and the question of whether or in which form these should be retained became a matter of controversy. In order that the natural sciences might be clustered at one location, the Humboldt University's Department of Mathematics and Natural Sciences approved plans in October 1991 for a natural sciences campus in Berlin-Adlershof (Humboldt-Universität zu Berlin 2011). In February 1995, the Humboldt University unveiled a master plan for the Adlershof campus, with which it hoped to meet the overall restructuring requirements and clarify the appointment of faculty members. Following an architectural competition, Volker Staab and Alfred Nieuwenhuizen were commissioned to design a new building in Adlershof, for use by the Institute of Chemistry of the Humboldt University. Since inauguration of the building in 2001, all of the Institute's departments of chemistry are gathered there under one roof.

Since the Adlershof design correlates the organisation of the Institute's various academic departments, the result is a highly permeable space both for teaching and research, which may as easily be read as a contemporary interpretation of Humboldt's educational ideal as The First Institute was back in 1900. One essential distinction is, however, that the teaching spaces (lecture halls, seminar rooms, labs for practical work) are now located in a separate central building used both by the physics and chemistry faculties. They thus afford students of different disciplines opportunity to mingle in the course of their everyday routines. Moreover, they are used by all of the departments on the campus for interdisciplinary or extra-curricular events, such as conferences, receptions and the like.

The now more stringent demands for laboratory safety and compliance with current regulations had a major influence on planning for the new building in Adlershof. The original plans for Hessische Strasse foresaw certain infrastructural safety precautions, such as fume hoods, fire prevention equipment and emergency showers, whereas those in the new institute in Adlershof maintain a strict separation of laboratory and office spaces, so as to reduce the potential health risk of laboratory chemicals. A further reason for creating such distinct spaces is the increasingly widespread use of highly sensitive measuring instruments that are viable only in certain lab environments, i.e. where factors such as temperature, air purity, oscillation and field stabilities can be controlled. The trend to greater specialisation within laboratories is clearly evident from the ground plan of the new building in Adlershof. The total floor area allotted to research laboratories requiring specific controlled environments has more than trebled while that reserved for lab-based practical work has been reduced by circa 10 per cent.

A further significant difference in the spatial planning for the new building is the creation of more office space. While office space in The First Institute on Hessische Strasse was originally limited to three consultation rooms measuring 20 square meters each, offices in the new building occupy more than 2,000 square meters and thus 30 per cent of the total floor area. This striking shift in the use of the building, from a predominantly lab-based to an office-based working environment, cannot be explained alone by the increased value now accruing to theoretical work. It also rests on the fact that experiments are increasingly conducted by digital and computer-aided means. The automation of technical procedures in lab work and likewise the use of digitally networked lab equipment enables experiments to be controlled and monitored from a distance, thus also from an office, with the result that physical lab-based activity is now generally limited to machine maintenance. Today the synthesiser systems increasingly deployed in the field of synthetic chemistry are able to carry out entire synthesis cascades, for example in the synthesis of sugar or peptides, with only minimal manual input. Even more complex experimental procedures rely ever more frequently on fully automated laboratory systems that are able to carry out chemical analyses as well as to document the results (Fraunhofer Institute 2012).

But not only experimentation is increasingly being moved from the lab to the office: of late, the library too follows the same trajectory. Above all, the shift from analogue to digital literature management has reduced the need to conduct research in a reading room.

	Analytical Chemistry	Inorganic Chemistry	Didactic Chemistry	Organic Chemistry	Physical Chemistry	Theoretical Chemistry	Institute of Chemistry	Total
Research Lab	334	1,238	91	1,397	646	---	53	3,759
Office	204	373	78	388	593	312	82	2,030
Practice Lab	54	341	58	---	---	98	1,120	1,671
Total	592	1,952	227	1,785	1,239	410	1,255	7,460

Table 3:
Allocation of floor area in m^2 according to the ground plan of the Emil Fischer Building, from: Archive of the Technical Department of the Humboldt University Berlin.

In contrast to the building on Hessische Strasse, no areas of the new building in Adlershof are explicitly earmarked as library space. The table above diagrammatically summarises the various areas of the building in terms of their use by respective user groups (**Table 3**).

These figures show that research laboratories are allotted more than twice as much floor area as the teaching laboratories. While the ratio was circa 1:1.5 when the former institute on Hessische Strasse was planned (**Table 1**), the reverse is now the case in the building currently in use (circa 1:0.4). One reason for this development is the growing use of sensitive laboratory apparatuses. Yet the shifting ratio of space devoted to research respectively to practical work also demonstrates the University's new understanding of its role in scientific research.

CONCLUSION

The study of the Institute of Chemistry at the Humboldt University shows that both tertiary education policy and changes in professional orientation have a decisive impact on the spatial organisation and use of the building. While in The First Institute the ratio of space devoted to research laboratories respectively to laboratories for practical work stood at 1:1.5, space distribution in the new institute building reverses the relationship. The total floor area allotted research laboratories has more than trebled while that for practical work—despite the two buildings' comparable capacity—has shrunk by circa 10 per cent.

Analysis of the curricula and spatial organisation facilitated insight into the evolution of tertiary chemistry studies in Berlin since 1900. With regard to the ratio of theoretical instruction respectively to lab-based practical instruction, it is striking that the number of hours devoted to the former has steadily increased since 1975, while the duration of lab-based practical instruction has remained almost constant, except for the period 1992–97, when new course regulations came into force. This implies an overall increase in course content of a theoretical nature and accordingly also in the amount of floor area made available for lectures, seminars and tutorials. A further change has been the growing need for office space. Around one-third of the total floor area of the building in Berlin-Adlershof is reserved for offices. These are increasingly used not exclusively for administrative or documentary procedures but as the *long arm* of the laboratories. Given the ongoing development and automation of chemical analysis procedures, one may surmise that the amount of time research scientists spend in the laboratory will shrink even further in the future. And there is another important factor to be considered when planning a contemporary laboratory: the new need to combine a variety of scientific disciplines under one roof. A case in point is the subject of the following chapter, the Humboldt University's latest natural sciences building project.

REFERENCES

Benner, Dietrich. 2003. *Wilhelm von Humboldts Bildungstheorie: Eine problemgeschichtliche Studie zum Begründungszusammenhang neuzeitlicher Bildungsreform*, 3rd ed., Weinheim: Juventa.

Fischer, Emil. 1900. *Eröffnungs-Feier des neuen I. Chemischen Instituts der Universität Berlin.* Berlin: Verlag von August Hirschwald.

Fischer, Emil and Max Guth. 1901. *Der Neubau des ersten Chemischen Instituts der Universität Berlin.* Berlin: Verlag von August Hirschwald.

Fraunhofer-Institut. 2012. "Vollautomatisches Labor der Zukunft." <http://www.fraunhofer.de/de/presse/presseinformationen/2012/februar/vollautomatisches-labor-der-zukunft.html> (accessed 16 March 2015).

Friedrich-Wilhelms-Universität Berlin. 1900. *Personal- und Vorlesungsverzeichnis*. University Library, Humboldt-Universität zu Berlin.

Friedrich-Wilhelms-Universität Berlin. 1920. *Personal- und Vorlesungsverzeichnis*. University Library, Humboldt-Universität zu Berlin.

Friedrich-Wilhelms-Universität Berlin. 1940. *Personal- und Vorlesungsverzeichnis*. University Library, Humboldt-Universität zu Berlin.

Herneck, Friedrich. 1976. "Zur Geschichte des Chemischen Instituts in der Hessischen Strasse." *Zeitschrift für Chemie* 16 (6): 209-214.

Humboldt-Universität zu Berlin. 1950. *Vorlesungsverzeichnis*. University Library, Humboldt-Universität zu Berlin.

Humboldt-Universität zu Berlin 1952. *Vorlesungsverzeichnis*. University Library, Humboldt-Universität zu Berlin.

Humboldt-Universität zu Berlin. 1967. *Vorlesungsverzeichnis*. University Library, Humboldt-Universität zu Berlin.

Humboldt-Universität zu Berlin. 1992. *Vorlesungsverzeichnis mit wissenschaftlichem Personalteil*. University Library, Humboldt-Universität zu Berlin.

Humboldt-Universität zu Berlin. 1997. *Vorlesungsverzeichnis mit wissenschaftlichem Personalteil*. University Library, Humboldt-Universität zu Berlin.

Humboldt-Universität zu Berlin. 2009. *Vorlesungsverzeichnis mit wissenschaftlichem Personalteil*. University Library, Humboldt-Universität zu Berlin.

Humboldt-Universität zu Berlin. "Biographie Fritz Haber." <http://www.sammlungen.hu-berlin.de/dokumente/9194> (accessed 27 March 2015).

Humboldt-Universität zu Berlin. 2011. "Chronologie des Umzugs nach Adlershof." Last modified July 27, 2011. <https://www.hu-berlin.de/ueberblick/campus/adlershof/ueberblick/geschichte/standardseite/umzug> (accessed 27 March 2015).

Meisel, Manfred. 2000. "100 Jahre Chemie in der Hessischen Strasse." In *Chemie an der Humboldt-Universität zu Berlin — 100 Jahre Chemische Institute in der Hessischen Strasse*. Festschrift, 10-19. Monsheim: Verlag für Marketing und Kommunikation.

Meisel, Manfred. 2011. "Die Chemischen Institute der Berliner Friedrich-Wilhelms-Universität und der Humboldt-Universität zu Berlin." *Humboldt-Spektrum* 18 (2-3): 18-23.

Meisel, Manfred and Ulrich Koert. 2000. "100 Jahre Chemisches Institut in der Hessischen Strasse in Berlin." *Nachrichten aus der Chemie* 48 (6): 772-77.

Ministerium für Hoch- und Fachschulwesen der Deutschen Demokratischen Republik. 1975. *Studienplan der Grundstudienrichtung Chemie zur*

Ausbildung an Universitäten und Hochschulen der DDR [Dossier DR 3]. Bundesarchiv für jüngere Deutsche Geschichte, Berlin.

Sauer, Joachim. 2000. "Das Institut für Chemie der Humboldt-Universität — Gegenwart und Zukunft." In *Chemie an der Humboldt-Universität zu Berlin — 100 Jahre Chemische Institute in der Hessischen Strasse*. Festschrift, 20-24. Monsheim: Verlag für Marketing und Kommunikation.

Thilo, Erich. 1953. "Bericht über die Wiederaufbauarbeiten am Chemischen Institut der Humboldt-Universität in den Jahren 1946-1954." Estate of Erich Thilo, Berlin.

Universität Berlin. 1946. *Vorlesungsverzeichnis*. University Library, Humboldt-Universität zu Berlin.

Zigan, Johanna. 2007. "*Der Erste Weltkrieg als Katalysator für die Akzeptanz der Ingenieurwissenschaften am Beispiel der RWTH Aachen.*" MA thesis, Rheinisch-Westfälische Technische Hochschule (RWTH) Aachen.

“The difference between European and Japanese labs is not the lab structure but the atmosphere. In Europe you have a long history of creating common spaces whereas we in Japan do not.”

Sathoshi Kera, Associate Professor,
Chiba University, Japan

8

PLANNING A NEW LABORATORY: THE INTEGRATIVE RESEARCH INSTITUTE FOR THE SCIENCES IN BERLIN-ADLERSHOF

Interview with Jürgen P. Rabe

You were directly involved in the proposal of the new building for the Integrative Research Institute for the Sciences (IRIS) at the Humboldt University Berlin, Adlershof campus. How did this come about?

The natural sciences at the Humboldt University Berlin moved to the Adlershof campus in 2003 and as a result were integrated into the largest and most modern science and technology park in all of Germany, which is currently made up of ten independent scientific facilities and some 500 companies. We wanted to take advantage of this unique environment in an innovative way. Christoph Markschies, the former president of the Humboldt University, encouraged me to develop, together with fellow colleagues, a concept to distinguish the natural sciences at the Adlershof campus. The result was a research programme, which would bring together leading experts in the fields of physics, chemistry, mathematics and computer sciences from the Humboldt University in an Integrative Research Institute for the Sciences. Its goal was to explore fundamentally novel hybrid materials and functional systems with previously inaccessible optical, electrical, mechanical and chemical characteristics and make them available for potential applications.

The main idea was based on my experience in the research division at the International Machines Corporation (IBM) during the mid-1980s. At the time IBM was the global leader in the field of information-storage technologies and for many years maintained a technological advantage over their competitors. To identify revolutionary technological innovations at the right time and implement them, IBM decided to conduct their own basic research to be able to detect early potentiality at the very limits of what was imaginable. One of IBM's maxim said, "it's not difficult to discover something new—difficult is *recognizing* when one has discovered something new". To achieve this, optimal conditions had to be created, and the IBM Almaden Research Center in San Jose, California was the result. The research centre was located not far from the production facility of magnetic discs, regarded the very best in the field worldwide at the time. There were highly esteemed experts in virtually every field of chemical physics as well as computer sciences with whom one could collaborate with at short notice. Here, at the outskirts of Silicon Valley cutting-edge research and development in the field of magnetic storage intersected. A similar interdisciplinary research association, open for application potentials is what we had in mind for IRIS.

The opportunity of obtaining substantial support for universities engaging in cutting-edge research became possible in the mid-2000s through the German government's *Excellence Initiative.* In addition, the president of the Humboldt University Berlin used the university's bicentennial in 2010—which also happened to be the centennial of the Kaiser Wilhelm Society for the Advancement of Science—as an opportunity to reinforce the university's commitment to cutting-edge research. These events set the stage for the emergence of a conceptual outline that would distinguish the natural sciences at Humboldt University. Unfortunately, the available research spaces at the campus in Adlershof were dedicated exclusively to separate subjects, each field had—as it does today—its own building, which in view of basic teaching is indeed perfectly reasonable. However, the idea behind IRIS would be virtually impossible to implement under these conditions, which meant that a new research building was necessary. Given the possibilities to receive funding by the German government for implementing large-scale research projects at universities distinguished by their outstanding scholarship and national importance, we submitted an application for IRIS that was approved in 2013.

Were there no interdisciplinary laboratories in Berlin prior to this?

The Physics Department of the University, which was built for Hermann von Helmholtz and went into operation in 1878, heralded the beginning of a functional architecture for science (**Fig. 8.1**) (Guttstadt 1886, 135–48). Coined the "palace of physics" by contemporaries, the department together with a second Chemistry Department and Technology Department was housed in a prestigious building that extended roughly 100 meters along the banks of the Spree, later called the Reichstagsufer, and which cost more than 1,5 million gold mark to build. Directly behind this building, located within the same complex, were the Physiology Department and Pharmacology Department (**Fig. 8.2**). This means: already by the end of the nineteenth century a building complex existed where the facilities of the natural sciences and life sciences were very close to each other, despite the fact that they were separate departments at this time. The integration of research and instruction

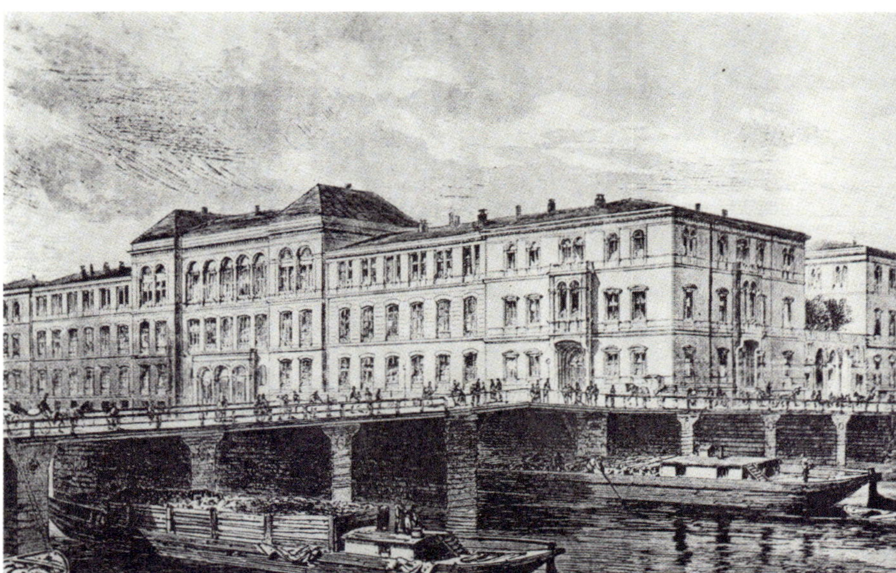

Fig. 8.1:
The Physics Department at Berlin University (1878), from: Auth and Kossack 1983: 567.

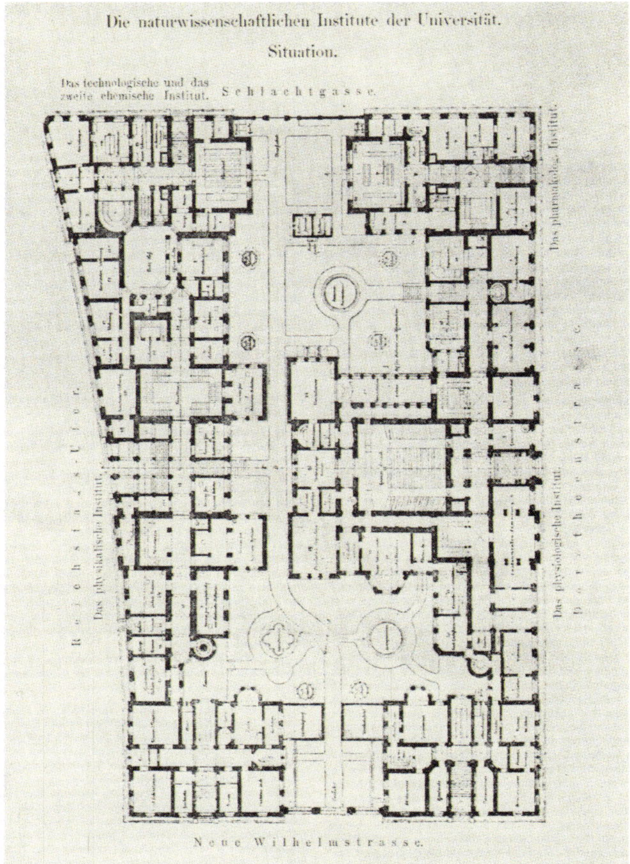

Fig. 8.2:
The Department of Natural Sciences at Berlin University, from: Guttstadt 1983: 570.

existed from early on at the University (See Chapter 7). However in contrast to today, the central inner city did not have access to nearby companies. Here lies one important difference between this nineteenth-century building and the envisaged IRIS research building in Berlin Adlershof.

After the Department of Physics on the Reichstagsufer was destroyed during the Second World War, a new departmental building for physics and electronics was erected in the 1980s in the Invalidenstraße 110 in the Berlin-Mitte district. At the beginning of the twenty-first century the various departments of natural science finally moved into the newly erected buildings in Adlershof (**Fig. 8.3**). The numerous architectural passageways in the Lise Meitner-Haus, the building housing the Department of Physics, already increased the potential for communication by greatly reducing dead-end hallways that would be used exclusively by staff working there (**Fig. 8.4**). Nevertheless, the special

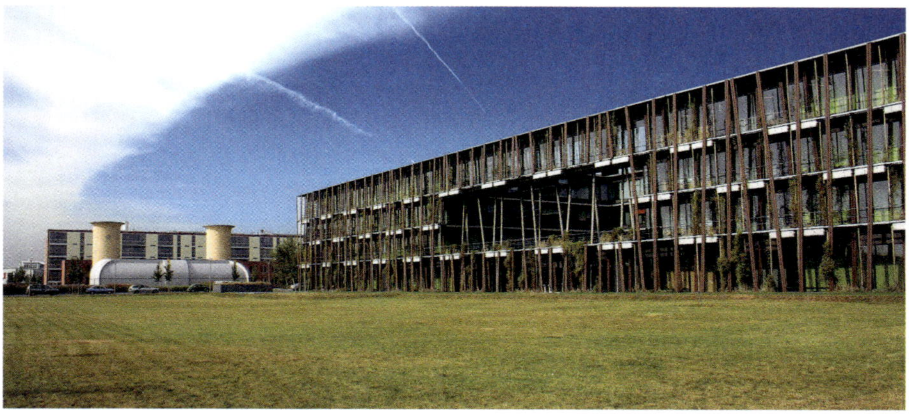

Fig. 8.3:
Augustin und Frank Architects, *Southern façade, Department of Physics*, Humboldt University, Berlin-Adlershof, 2003. Photo © Heike Baeckmann

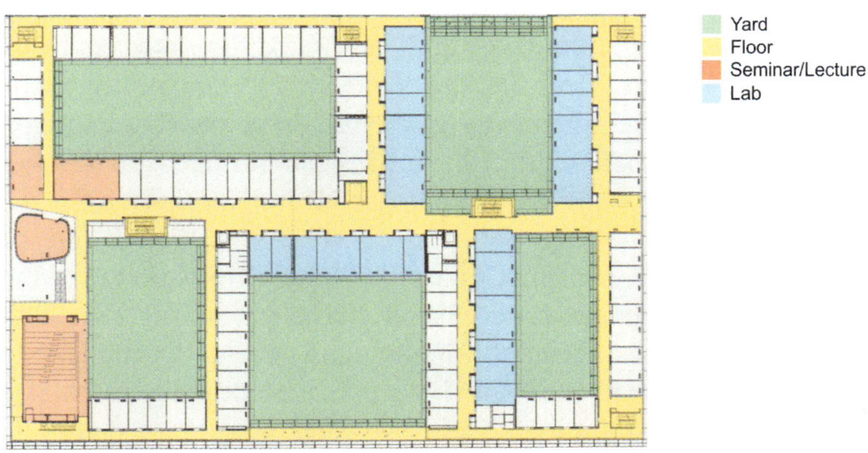

Fig. 8.4:
Augustin und Frank Architects, *First floor ground plan, Department of Physics*, Humboldt University, Berlin-Adlershof, 2003. Design © Augustin und Frank

laboratories for a Materials Science Centre planned for this site were not approved by the state government, which means such interdisciplinary research will only be feasible in the planed IRIS research building.

Are there any contemporary interdisciplinary laboratories that served as models for IRIS?

From my own research I was already familiar with a few laboratories distinguished by their interdisciplinary work, such as the IBM Almaden Research Center that I have already mentioned, the Max Planck Institute for Polymer Research in Mainz, as well as the Max Planck Institute for Colloids and Interfaces in Potsdam. Furthermore, I had the unique opportunity together with the authors of this book—a team made up of physicists, chemists, architects and art historians from the Excellence Cluster *Image Knowledge Gestaltung. An Interdisciplinary Laboratory*—to visit a whole range of new laboratories in Europe, Asia and the United States where outstanding interdisciplinary research is being conducted. From each example one could draw different conclusions; some very concrete, others much more general.

We learned from the Fritz Haber Institute of the Max Planck Society (under construction since 2010) located in Berlin, that with a decentralized temperature control one could achieve extremely stable temperatures, even in very large laboratories. However, it also became clear that the choice of a free-spanning ceiling constrained the dimensions of the lab, since the cost increase rapidly with the span width. This influenced our decision making process when proposing the dimensions for larger laboratories in the IRIS research building.

At the NanoTecCenter Weiz (2007) in Austria we encountered a very compact, functional and efficient facility, which perhaps came closest in satisfying the requirements for the IRIS research building, and from which we could adopt many ideas, specifically concerning the design of clean room laboratory spaces.

The MESA+ Institute for Nanotechnology (2010) at the University of Twente in the Netherlands showed the positive influence a more open architectonic structure can have. Here the glazed visitor gallery around the entire clean room complex enables a spectacular open access to research, something we also envisioned for the IRIS research building. This feeling of openness achieved through the use of glass and an array of visual perspectives continues throughout the MESA+ (See Chapter 4). In addition, the close connection between laboratories and the office workspaces of colleagues coming from different disciplines is perceived to have a predominantly positive influence on interdisciplinary cooperation (See Chapter 6). The necessary flexibility of the laboratory facility was achieved at MESA+ through a modular supply system, of course something that comes at a price.

The facilities at the Chiba University in Tokyo, in contrast, were generally comprised of austere single laboratories built during the later half of the twentieth century. The spatial relationship and the communicative connection between the individual laboratories did not seem to be a priority in the architectural planning. The members of our interdisciplinary team assessed this particular point differently. The architects and art historians were critical of this design, whereas the scientists, who actually worked in these spaces, positively noted that the strict tailoring of each individual room and its respective function eased the ability to concentrate on the task at hand. This shows that even plain and functional laboratory spaces have their merits. The communication amongst the working groups simply takes place elsewhere. At Chiba University there was instead a large common room where colleagues can cook together and host social events.

The Centre for Advanced 2D Materials and Graphene Research Centre (2010) at the National University of Singapore is an interdisciplinary facility that places special focus on modes of communication. The relatively barebones space had been made more communication-friendly through the incorporation of certain architectural elements such as inscribable glass doors and common spaces. In practice it was then determined which of these elements were helpful and which were superfluous. This experimental approach with spatial configurations made clear to us that a certain degree of flexibility is desirable in the design of laboratories' interior spaces.

In the United States we visited the Frick Chemistry Laboratory (2010) at Princeton University (See Chapters 5 and 6), a laboratory where chemical research is brought together with other disciplines and close interdisciplinary collaboration is encouraged. Based on my own research-related experiences in this building, I can confirm that the building indeed enables interdisciplinary collaboration and can serve as an example for other laboratories. The building's open and expansive architecture is certainly a significant element of the laboratory's overall success. The building contains a large central light-drenched atrium, through which the laboratories and offices—all distinguished by a well thought-out design—are accessible, regularly offering pleasant unobstructed views into the open (See Chapter 6). Nonetheless, this architectural structure also has its disadvantages as some users noted the limited room for storage as well as a lack of privacy and flexible space.

We also visited other contemporary interdisciplinary laboratories with different research objectives. The Wyss Institute for Biologically Inspired Engineering (2009) at Harvard University is housed in a building originally intended for other purposes and was later modified for its current use. Here, the flexibility of fixtures and the completely adjustable laboratory benches including the utility supply of gas and water stood out as particularly noteworthy features. Finally, the MIT Media Lab (2009) in Boston demonstrated the stimulating influence a building's architecture can have on the indoor atmosphere by intentionally allowing unusual viewing perspectives throughout the building (See Chapter 6).

Today laboratory architecture usually follows established standards. Does an interdisciplinary laboratory lead to the dissolution of conventional laboratories?

Though the IRIS research building follows the principles of interdisciplinarity, there will still be a whole range of traditionally specialised laboratories available on site (**Fig. 8.5**). Laboratories for work on wet chemistry or biology are outfitted with the corresponding safety precautions, whereas optical laboratories require a special control of temperature stability and dimming options. Laboratories for high-resolution microscopic techniques present different challenges for laboratory spaces, given that the procedures must be conducted free from vibrations and shielded against electromagnetic fields. In such cases adhering to conven-

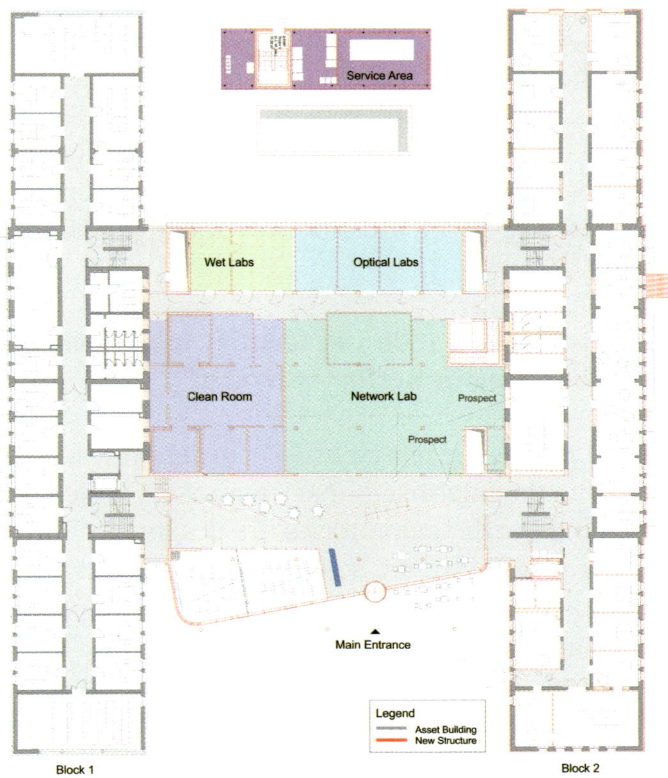

Fig. 8.5:
Nickl & Partner Architects, *Floor plan, ground floor*, IRIS, Berlin-Adlershof. Design © Nickl & Partner Architekten AG

tional specialised laboratories is necessary. At the same time IRIS will also have large-scale laboratories commonly seen at bigger research institutions. These types of laboratories enable the set up of a wide range of larger equipment, and thereby increase the exchange between various groups working in close proximity. In addition there will be specialised jointly used laboratories, where specific procedural steps will be seamlessly intertwined, which until now are conducted separately not only in different spaces but sometimes even at completely different institutions. This includes, the connection of wet chemistry production processes in a dust-free environment with stable air humidity on the one hand, and ultra-high vacuum based processes on the other hand. Depending on the processes, these joint laboratories will also operate with elaborate online-methods of analysis. In the planning of the joint laboratory in the IRIS research building first a requirement profile had to be drafted to ensure the versatility of future demands. Finally, there will be laboratories used for the production of prototypical objects that can be operated in joint cooperation with companies interested in the broader uses and applications of the findings.

What innovative potential do you identify in the design of this new research building in Berlin-Adlershof?

First of all, the planned research itself is geared towards innovation, and a key is the interdisciplinary approach, which creates new demands in how the laboratories are designed. Since research at IRIS no longer recognizes disciplinary boundaries, joint laboratories as well as versatile and flexible spaces are particularly suitable in accommodating intensive collaboration. This is the primary difference between the new research building for IRIS and disciplinary-based institutional buildings where different types of laboratories are traditionally separated. In addition to the collaboration in various larger groups, we also wanted to place emphasis on the ability of individual researchers to withdraw to private spaces, something that is just as important for innovation.

How was the planning process for this new building carried out? Who was involved and at what point?

Once the application for funding was approved in 2013, a small work group supported by about a dozen colleagues who were part of the research program began drafting a detailed program of requirements. Based on this program the relevant state

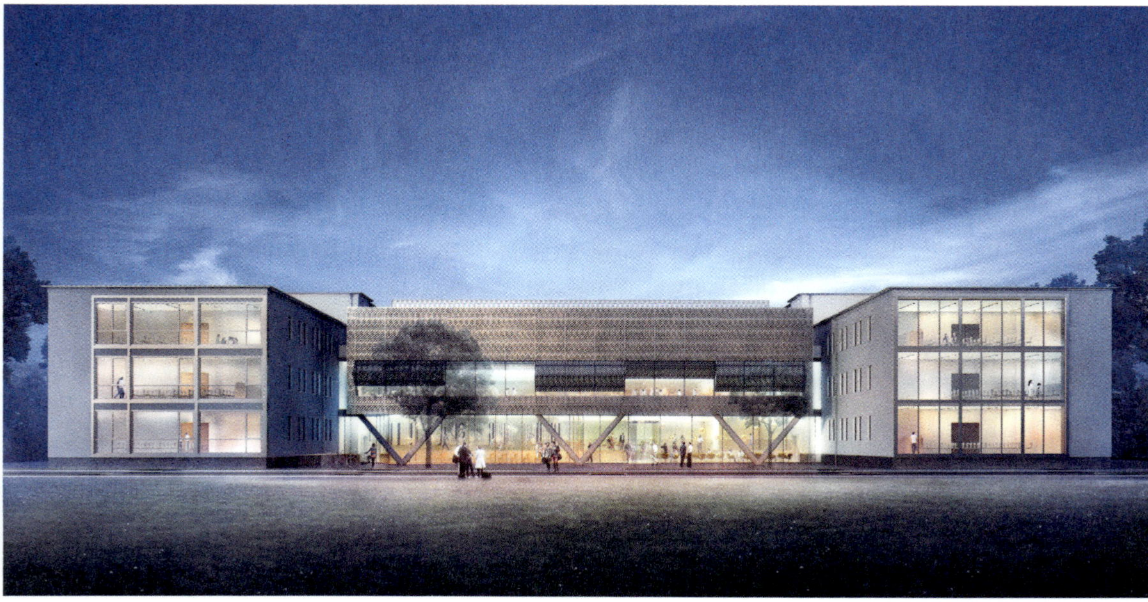

Fig. 8.6:
Nickl & Partner Architects, *IRIS*, Berlin-Adlershof. Design © Nickl & Partner Architekten AG

administration announced a limited call for architectural proposals. The design submitted by Nickl & Partner Architects was declared the winner (**Fig. 8.6**). The detailed plans were finalized in the summer of 2015, then certified by the relevant state administration and followed by the bidding process for the various construction firms. Construction is planed to commence during the first half of 2016, with completion scheduled in 2018.

Users' involvement in the planning process of research buildings varies from country to country. How much influence did you have on the building's design?

Our influence on the building's design was considerable. On the basis of our proposed research program the Humboldt University offered a suitable site of land for our use, which happened to include two buildings in need of renovation. The call for architectural proposals was also open to designs that envisaged the demolition of one or more of these buildings. The winner of the competition proposed keeping the existing buildings and integrating them into the overall design. This proposed connection of two separate construction projects was only possible due to the positive cooperation of those involved. The only bad news was that the total costs initially proposed had been grossly underestimated. This meant that the interweaving of theoretical and experimental research groups in combined workspaces and the enhancement of communication through visual perspectives throughout the building could only be conditionally guaranteed.

Who moderated the users' interests and how where they communicated to the architects?

The users' feedback was gathered by the IRIS office and from there was communicated to the various administrative bodies and designers, which included the Construction Department of the Humboldt University, the State Department of Education, Youth and Academics, the Office of Urban Development and Environment, which is in charge of construction, as well as the architects at Nickl & Partner. In the end the contractor of course makes the decisions, but generally the input of the users was duly considered throughout the planning process.

Which of your ideas were carried out, which remained unfulfilled?

Our central concerns such as opening the call for proposals to architectural solutions that were not constrained by the existing structures, or our desire for a direct connection between the research building and the existing structures were met. Our numerous requests for connections between individual and special laboratories were also satisfied to the greatest extent possible. Unfortunately the interconnection between office and laboratory spaces was restricted on financial grounds, which in the end meant that the office spaces would be housed in one of the existing structures and the laboratories in the newly constructed wing.

Imagine an ideal world, where financial considerations and regulated planning processes did not play a role. How would the Integrated Research Institute for the Sciences be different?

Perhaps the spaces would be a little more integrative, as seen at the MIT Media Lab; maybe a little bit more flexible, as implemented at the laboratory facilities at MESA+ or Wyss Institute; and maybe a little bit more open like the architecture of MESA+ or the Frick Chemistry Building. But for all intents and purposes we are very excited about the opening of IRIS as it is currently planed.

(Translation by Landon Little, Berlin)

REFERENCES

Auth, Joachim and Heinz Kossack: "Zur Lage der Physik an der Berliner Universität vor der Errichtung des Instituts am Reichstagufer." 1983. *Wissenschaftliche Zeitschrift der Humboldt-Universität zu Berlin.* Mathematisch-Naturwissenschaftliche Reihe 32 (5): 557-567.

Guttstadt, Albert "Das physikalische Institut (NW. Neue Wilhelmstrasse 16a)." 1886. *Wissenschaftliche Zeitschrift der Humboldt-Universität zu Berlin.* Mathematisch-Naturwissenschaftliche Reihe 32: 5 (1983): 569-76. Orig. publ. *Die naturwissenschaftlichen und medicinischen Staatsanstalten Berlins.* Berlin: Verlag von August Hirschwald, S. 135-48.

Schmidt, Marco and Thies Schröder. 2003. "Physikinstitut in Berlin" *Baumeister. Zeitschrift für Architektur* 100 (2): 66-68.

NOTES ON CONTRIBUTORS

RONALD GÖBEL

studied chemistry at the University of Potsdam, from which he graduated in 2007. In 2011 he received his PhD for research in Hybrid Materials of Mesoporous Silica and Ionic Liquids. His research focus is the field of nanostructured materials and systems. From 2013 to 2014 he was a research associate at the Cluster of Excellence *Image Knowledge Gestaltung* at the Humboldt University Berlin. He works as a research assistant at the Deutsches Museum in Munich.

SABINE HANSMANN

studied architecture at the Academy of Arts and Design in Stuttgart and the University of the Arts in Tokyo. After her graduation in 2012, she worked in several architectural offices and as a freelance editor for the architectural journal *ARCH+*. She is currently a research associate at the Cluster of Excellence *Image Knowledge Gestaltung* at the Humboldt University Berlin and is completing her PhD thesis on monospace as an area of opportunities.

CHARLOTTE KLONK

is Professor of Art History and New Media at the Institute of Art and Visual History and Principal Investigator at the Cluster of Excellence *Image Knowledge Gestaltung* at the Humboldt University Berlin. Previously, she was a research fellow at Christ Church Oxford and Lecturer in the History of Art Department at the University of Warwick. She has been a fellow at the Max Planck Institute for the History of Science, at the Institute of Advanced Studies in Berlin and at the Clark Art Institute in Williamstown, Massachusetts. Her publications include *Science and the Perception of Nature* (Yale University Press, 1998), *Spaces of Experience: Art Gallery Interiors from 1800 to 2000* (Yale University Press, 2009) and, with Michael Hatt, *Art History: A Critical Introduction to its Methods* (Manchester University Press, 2005). Most recently, she has co-edited with Jens Eder, *Image Operations: Visual Media and Political Conflict* (Manchester University Press, 2016).

CHRISTINA LANDBRECHT

studied art history at the Humboldt University Berlin, from which she graduated in 2009. From 2009 to 2012 she worked as an assistant curator at the modern art museum Berlinische Galerie. She is a research associate at the Cluster of Excellence *Image Knowledge Gestaltung* at the Humboldt University Berlin and is currently completing her PhD thesis on "The Laboratory as Operation Manual: Experimental Science and Contemporary Art." She also works as a freelance curator and critic, contributing regularly to *radioeins and Deutschlandradio Kultur.*

HENRIKE RABE
graduated in architecture from the Technical University Berlin in 2007. She worked as an architect at Brisac Gonzalez Architects in London and, from 2009 to 2012, as Senior Architect at Kazuhiro Kojima + Kazuko Akamatsu/CAt in Tokyo, where she was in charge of the planning of universities, schools, museums and libraries. She is a research associate at the Cluster of Excellence *Image Knowledge Gestaltung* at the Humboldt University Berlin and currently completing her PhD thesis on "Architectures of Knowledge." She also writes regularly for *Bauwelt* and other architectural magazines.

JÜRGEN P. RABE
is Professor of Physics of Macromolecules in the Department of Physics and Associate Investigator at the Cluster of Excellence *Image Knowledge Gestaltung* at the Humboldt University Berlin. His research investigates the correlation between the structure and dynamics of molecular systems and interfaces with mechanical, electronic, optical and (bio-)chemical characteristics, from molecular to macroscopical length and time scales. He is Founding Chairman of the *Integrative Research Institute for the Sciences — IRIS Adlershof*.

VERENA STRAUB
studied art history at the Humboldt University Berlin, from which she graduated in 2012. She is a research associate at the Cluster of Excellence *Image Knowledge Gestaltung* at the Humboldt University Berlin and at the research institution *Affective Societies* at Free University Berlin. She is currently completing her PhD thesis on video testimonies of suicide bombers and their adaptation in contemporary art. She also works as a freelance journalist, contributing to national daily papers and art magazines.

ACKNOWLEDGEMENTS

This publication is part of a research project financed by the Cluster of Excellence *Image Knowledge Gestaltung* at the Humboldt University in Berlin from 2012 to 2016. The editor and authors are very grateful to the Cluster for their longstanding support, advice and generous financial aid. Our special thanks go to the English translators, Jill Denton (Berlin) and Landon Little (Berlin), whose efforts went well beyond the call of duty. We also thank the designer Katja Gretzinger for her creative suggestions. Many thanks too go to our student assistants Venetsiya Dimitrova, Laura Enghusen, Alisa Laila Fluhrer, Cem Kozcuer, Maren Krause, Patryk Kujawa and Catherine Slusher who provided practical assistance and helped to prepare and evaluate data.

The project owes more than is reflected in the book to Prof. Jurriaan Huskens, Miriam Luizink, Gerard Roelofs, Ruud van Leeuwen, and Eddy Ruiter from MESA+ at the University of Twente (The Netherlands); Henk Heidekamp from the ERIBA at the University Medical Center Groningen (The Netherlands); Prof. Emil List-Kratochvil, Dr. Andreas Klug, and Dr. Stefan Sax from NanoTecCenter Weiz (Austria); Dr. Ayis Antoniou and Dr. Mary Tolikas from the Wyss Institute at Harvard University (USA); Prof. Steven L. Bernasek, Prof. Antoine Kahn, and Prof. Jeffrey Schwartz at Princeton University (USA); Joseph Mannino at Columbia University (USA); Prof. Nobuo Ueno and Prof. Sathoshi Kera at Chiba University (Japan); and to Prof. Barbaros Özyilmaz, Prof. Christian Kurtsiefer, Prof. Antonio Castro Neto, Prof. Andrew Wee, and Prof. Chen Wei at the National University Singapore (Singapore). All went out of their way to give us access to their laboratories and discuss with us in depth his or her experiences and expectations. Our colleague Prof. Norbert Koch, Humboldt University Berlin, was instrumental in getting the project started and remained a guiding spirit throughout. Prof. Finn Geipel, Technical University Berlin, also part of the team, provided invaluable insights into laboratory architecture and was instrumental in shaping the spatial and functional analyses. Lastly, our thanks must go to our publisher, Walter de Gruyter, and to Dr. Katja Richter, our editor, for efficient, intelligent, and professional guidance throughout the publishing process.

INDEX